John Bernhard Stallo

The Concepts and Theories of Modern Physics

Elibron Classics
www.elibron.com

Elibron Classics series.

ISBN 1-4021-9707-1 (paperback)
ISBN 1-4021-2232-2 (hardcover)

This Elibron Classics Replica Edition is an unabridged facsimile of the edition published in 1900 by Kegan Paul, Trench, Trübner & Co., Ltd., London.

XLIV. **MAN BEFORE METALS.** By N. JOLY, Correspondent de l'Institut de France. Fifth Edition. With 148 Illustrations.

XLV. **The SCIENCE of POLITICS.** By Prof. SHELDON AMOS. Third Edition.

XLVI. **ELEMENTARY METEOROLOGY.** By ROBERT H. SCOTT. With 11 Plates and 40 Figures in Text. Seventh Edition.

XLVII. **The ORGANS of SPEECH.** By GEORG HERMANN VON MEYER. With 47 Illustrations.

XLVIII. **FALLACIES**: a View of Logic from the Practical Side. By ALFRED SIDGWICK. Second Edition.

XLIX. **The ORIGIN of CULTIVATED PLANTS.** By ALPHONSE DE CANDOLLE. Second Edition.

L. **JELLY FISH, STAR FISH, and SEA URCHINS.** Being a Research on Primitive Nervous Systems. By G. J. ROMANES, LL.D., F.R.S. Second Edition.

LI. **The COMMON SENSE of the EXACT SCIENCES.** By the late WILLIAM KINGDON CLIFFORD. Third Edition. With 100 Figures.

LII. **PHYSICAL EXPRESSION**: its Modes and Principles. By FRANCIS WARNER, M.D., F.R.C.P. Second Edition. With 50 Illustrations.

LIII. **ANTHROPOID APES.** By ROBERT HARTMANN. With 63 Illustrations. Second Edition.

LIV. **The MAMMALIA in their RELATION to PRIMEVAL TIMES.** By OSCAR SCHMIDT. Second Edition. With 51 Woodcuts.

LV. **COMPARATIVE LITERATURE.** By H. MACAULAY POSNETT, LL.D.

LVI. **EARTHQUAKES and other EARTH MOVEMENTS.** By Prof. JOHN MILNE. With 38 Figures. Fourth Edition, revised.

LVII. **MICROBES, FERMENTS, and MOULDS.** By E. L. TROUESSART. With 107 Illustrations. Third Edition.

LVIII. **GEOGRAPHICAL and GEOLOGICAL DISTRIBUTION of ANIMALS.** By Prof. A. HEILPRIN. Second Edition.

LIX. **WEATHER**: a Popular Exposition of the Nature of Weather Changes from Day to Day. By the Hon. RALPH ABERCROMBY. With 96 Figures. Fourth Edition.

LX. **ANIMAL MAGNETISM.** By ALFRED BINET and CHARLES FÉRÉ. Fourth Edition.

LXI. **MANUAL of BRITISH DISCOMYCETES**, with descriptions of all the Species of Fungi hitherto found in Britain included in the Family, and Illustrations of the Genera. By WILLIAM PHILLIPS, F.L.S. Second Edition.

LXII. **INTERNATIONAL LAW.** With Materials for a Code of International Law. By Professor LEONE LEVI.

LXIII. **The GEOLOGICAL HISTORY of PLANTS.** By Sir J. WILLIAM DAWSON. With 80 Illustrations.

LXIV. **The ORIGIN of FLORAL STRUCTURES THROUGH INSECT and other AGENCIES.** By Prof. G. HENSLOW. Second Edition.

LXV. **On the SENSES, INSTINCTS, and INTELLIGENCE of ANIMALS, with special reference to INSECTS.** By Sir JOHN LUBBOCK, Bart., M.P. With 118 Illustrations. Third Edition.

LXVI. **The PRIMITIVE FAMILY in its ORIGIN and DEVELOPMENT.** By C. N. STARCKE. Second Edition.

THE

INTERNATIONAL SCIENTIFIC SERIES

VOL. XLII.

"*Inficitur autem intellectus humanus ex intuitu eorum, quae in artibus mechanicis fiunt, in quibus corpora per compositiones aut separationes utplurimum alterantur; ut cogitet, simile quiddam etiam in naturâ rerum universali fieri. Unde fluxit commentum illud elementorum deque illorum concursu, ad constituenda corpora naturalia.*"

BACO VERUL., NOV. ORG., LIB. I, APH. 66.

THE

CONCEPTS AND THEORIES

OF

MODERN PHYSICS

BY

J. B. STALLO

FOURTH EDITION

LONDON

KEGAN PAUL, TRENCH, TRÜBNER & CO. LTD.

PATERNOSTER HOUSE, CHARING CROSS ROAD

1900

To the Memory

OF

DR. WALTER STALLO.

. . . . . . *summa tamen omnia constant.*

INTRODUCTION

AND PREFACE TO THE SECOND EDITION.

The present new edition of this volume is substantially a reprint of the first. I have made a brief addition to the fifteenth chapter, taking note of G. H. Darwin's investigations respecting the influence of the solar tides upon the rotation of the planets, as bearing upon the anomalies observed in the periods of revolution of the satellites of Mars; but beyond this I have confined myself to the correction of a number of typographical errors, and of a few verbal inaccuracies. It was at one time my intention to expand some of the early chapters, by extending the considerations therein presented to other physical and chemical details, and to supplement the ninth, tenth, eleventh, and twelfth chapters, by a discussion of certain general topics, such as the concepts of *substance* and *cause* as commonly dealt with in physical science. But I found that it was impracticable to carry this intention into effect without transcending the limits, if not of my subject, at least of the space at my command. And there being nothing in the English and American criticisms of the book, so far as they have fallen under my notice, which, in my judgment, requires a serious revision of any part of its contents, I have concluded to republish it in its original form. Nevertheless, I avail myself of this opportunity to reply to some of the criticisms alluded to, chiefly to the end of removing a curious misapprehension which is common to nearly all my critics, but incidentally also of meeting certain objections that are more or less founded upon or connected with it.

I.

The misapprehension I speak of is very surprising, in view of the explicit declaration, contained in the very first sentence of my preface, that the book is "designed as a contribution not to physics, nor certainly to metaphysics, but to the theory of cognition." Notwithstanding this declaration, most of my critics assume it to be my purpose to expose the shortcomings and defects of particular physical theories as devices for the colligation of facts, or as instruments of research, and suppose that my endeavor is simply, as one of my critics expresses it, "to pick flaws in these theories," or, in the language of another critic, "to crassify and develop contradictions" between them, to "set facts by the ears," and "bump friendly heads together"—in short, in the spirit of a sort of scientific pyrrhonism, to discredit the familiar methods of physical science, if not to invalidate its results. And they complain that I fail to apprehend what one of them is pleased to term the "laboratory function" of a physical theory or hypothesis, and to appreciate the distinction between a "working hypothesis" and a theory advanced with the claim of its final validity or truth.

Now, the fact is that, for the purposes of the inquiry to which my book is devoted, I am not *directly* concerned with the "laboratory function" of "working hypotheses" or physical theories at all. My object is to consider current physical theories and the assumptions which underlie them in the light of the modern theory of cognition—a theory which has taken its rise in very recent times, and is founded upon the investigation, by scientific methods analogous to those employed in the physical sciences, of the laws governing the evolution of thought and speech. Among the important truths developed by the sciences of comparative linguistics and psychology are such as these: that the thoughts of men at any particular period are limited and controlled by the forms of their expression, viz., by language (using this term in its most comprehensive sense); that the language spoken and "thought in" by a given generation is to a certain extent a record of the intellectual activity of preceding generations, and thus embodies and serves to perpetuate its errors as well as its truths; that this is the fact hinted at, if not accurately expressed, in the old observation according to which every distinct form or system of speech involves a distinct metaphysical theory; that the metaphysical systems in vogue at any particular epoch, despite their apparent differences and antagonisms, on proper analysis are found to be characterized by certain common features in which the latent metaphysics of the language in which such systems have originated, or are presented, are brought to view; that philosophers as well as ordinary men are subject to the thralldom of the intellectual prepossessions embodied in their speech as well as in the other inherited forms of their mental and physical organizations, and are unable to emancipate themselves from this thralldom otherwise than by slow and gradual advances, in conformity to the law of continuity which governs all processes of evolution whatever. It being my belief that all this applies to the votaries of science as well as to the devotees of metaphysics or ontology, I sought to enforce this belief by an examination of the general concepts and theories of modern physics. According to the opinion of contemporary men of science, these concepts and theories are simply generalizations of the data of experience, and are thus not only independent of the old *a priori* notions of metaphysics, but destructive of them. But, although the founders of modern physical science at the outset of their labors were animated by a spirit of declared hostility to the teachings of mediæval scholasticism—a fact which is nowhere more conspicuous than in the writings of Descartes—nevertheless, when they entered upon the theoretical discussion of the results of their experiments and observations, they unconsciously proceeded upon the old assumptions of the very ontology which they openly repudiated. That ontology—founded upon the inveterate habit of searching for "essences" by the interpretation of words and the analysis of the concepts underlying them, before the relations of words to thoughts and of thoughts to things were properly understood—was characterized by three great errors: its hypostasis of concepts (notwithstanding the protest of the nominalists against the reification of universals); its disregard of the twofold relativity of all physical phenomena; and its confusion of the order of intellectual apprehension with the order of nature. These errors gave rise to a number of cardinal doctrines respecting the "substance of things," among which were the assertion of its existence as a distinct thing or real entity, apart from its properties; the further assertion of its absolute permanence and immutability; and, finally, the assertion of the absolute solidity and inertia of its parts and their incapacity to act upon each other otherwise than by contact.* And all these doctrines lie at the base, not only of Cartesian physics and metaphysics, but of the scientific creed of the great majority of the physicists of the present day. The eminent physicist and physiologist who declares that "before the differential equations of the world-formula could be formed" [i. e., before the ultimate, true, and exhaustive theory of the universe could be constructed], "all processes of nature must be reduced to the motions of a substratum substantially homogeneous, and therefore totally destitute of quality, of that which appears to us as heterogeneous matter—in other words, all quality must be explained by the arrangement and motion of such a substratum," † and the equally distinguished physicist and mathematician who enters upon the attempt at a solution of the problem thus stated by endeavoring to deduce the phenomenal diversities and changes of the universe from imaginary vortical motions of the undistinguishable parts of an assumed universal, homogeneous, continuous and incompressible fluid, are both as truly instinct with the spirit of the old *scientia entis quatenus entis* as the most ardent disciple of the Stagirite in the times of Erigena or Aquinas. The physicist who insists upon impact theories of gravitation, cohesion, or chemical affinity,

* In this connection it may be worth while to direct the attention of our modern "Baconian" physicists to the fact that the proposition, according to which there can be no physical action without contact, is one of the fundamental doctrines of Aristotle. See the references in *Zeller's Philosophie der Griechen, second ed., II_2, p. 268.*

† "Ehe die Differentialgleichungen der Weltformel angesetzt werden koennten, muessten alle Naturvorgaenge auf Bewegungen eines substantiell unterschiedslosen, mithin eigenschaftslosen, Substrates dessen zurueckgefuehrt sein, was uns als verschiedenartige Materie erscheint, mit andern Worten, alle Qualitaet muesste aus Anordnung und Bewegung solchen Substrates erklaert sein."—*Du Bois-Reymond, Ueber die Grenzen des Naturerkennens, 2. ed., p. 5.*

has the same intellectual blood in his veins which coursed in those of the old disputants about "first matter" or "substantial forms." When the Professor of Physics in the University of Edinburgh teaches that matter is absolutely passive, *dead*,* that all physical action is action by contact, that nothing is real which is not indestructible,† etc., he stands as unmistakably upon scholastic ontological ground as did Descartes or any of his ecclesiastical contemporaries. The proposition of the modern kinematist, that the true explanation of the phenomena of heat, light, electricity, magnetism, etc., consists in their reduction to the elements of matter and motion, differs in little else than its phraseology from the metaphysical theorem that all the "secondary qualities" of the universal substance are mere specifications or derivatives of its "primary qualities."

Such being the theme of the little book which is now before the reader, it is readily seen that it became incumbent on me to establish two main propositions. The first of these is the proposition just stated and briefly illustrated, that the general principles of the atomo-mechanical theory, which is said to be the basis of modern physics, are substantially identical with the cardinal doctrines of ontological metaphysics; and the second, that the fundamental errors of ontology become apparent in proportion to the advance of physical science, inasmuch as the four great assumptions which the atomo-mechanical theory necessarily involves (viz., those of the absolute equality, inertia, and rigidity of the elementary atoms and of the essentially kinetic character of all physical energy) are distinctly and irreconcilably

* *Cf. The Unseen Universe*, § *104; Tait, On some Recent Advances in Physical Science, p. 24, etc.*

† *On Some Recent Advances, etc., pp. 14–17.*

at war with the widest and most trustworthy experiential inductions of physics, chemistry, and astronomy. In the natural order of my discussion, the establishment of the first proposition would have preceded that of the second ; and, when I first undertook to write the book, I actually began to conform to that order by a brief review of the history of physics from the time of Descartes. A very cursory examination of the *Principles*, the *Discourse on Method*, and parts of the *Meteors*, of Descartes, is sufficient to show that, notwithstanding his own sincere belief that his philosophical and scientific doctrines were thoroughly subversive of Aristotelianism and scholasticism, he was at bottom as thorough a scholastic ontologist as the doctors of the Sorbonne whose favors and indulgence he openly courted while he privately despised them. And an equally cursory glance at the history of theoretical physics since Descartes's day reveals the fact that the persistent denial of the possibility of *actio in distans* and the incessant renewal of the attempts at kinetic explanations of gravity, cohesion, affinity, electric and magnetic attraction, etc., are neither more nor less than recrudescences of ontological Cartesianism. In the progress of my work I became apprehensive, however, lest my scientific readers should be deterred from perusing my pages by the quasi-metaphysical aspect of the introductory parts. So I concluded to reverse the order of the argument, and to present the second propositon first. It is not improbable that this was unwise. For, on the one hand, it has led to the strange consequence that the task of reviewing the book has generally been assigned to specialists, who not only abhor metaphysics, but regard as metaphysical everything which does not present itself in the guise of a differential equation or of an atomic formula

—so that the editor of *Mind*, for instance, has enthroned in the critical chair a learned gentleman who declares that he does not venture an opinion on what he terms "the more strictly metaphysical part" of the book, and the whole tenor of whose discussion shows that he is unaware even of the existence of the science to which I attempted to make a contribution, although this science is repeatedly referred to in the very number of *Mind*, in which his criticism appears, under the name of *epistemology;* and, on the other hand, it has brought upon me the charge of being a mere destructionist who fails to understand the laboratory functions of physical theories, and who is intent upon subverting these theories without offering or suggesting others to replace them.

If the foregoing exposition has properly served to define the purpose and scope of my inquiry, the reader will have no difficulty in seeing that the strictures of my critics upon my supposed ignorance of the distinction between working hypotheses and ultimate theories, and of the "radically different tests of logical canonicity" to be applied to them, as well as their animadversions upon my inordinate stickling for "exiguous consistency" between the parts of a theory in preference to "its flexile and serviceable adaptability to facts of many and diverse orders," etc., etc., are as irrelevant as they are unfounded. The tone in which these strictures are presented, as though they were in any sense refutations of the general argument of my book, affords curious evidence of the confusion which runs riot among the theoretical speculations of modern physicists. For purposes of practical scientific research the chief value of hypotheses or provisional theories lies, no doubt, in their capacity to effect a momentary fusion of experimental results; and the physicist or chemist in his laboratory may, with a certain degree of safety, either wholly ignore the question of their ultimate validity, or content himself with placing it in the distant perspective. The primary relation of these hypotheses or theories to scientific research is analogous to that of language to the ordinary operations of thought: they serve to fix and record the results of experiment and observation. But, generally speaking, hypotheses are more than mere arbitrary and artificial devices for the enchainment and classification of facts. They are in most cases guesses at the ultimate truth suggested by the analogies of experience, and are primarily used as working hypotheses only in the sense that they afford a basis for further experiment and observation whereby their ultimate validity is to be established or overthrown. In the progress of the various attempts at their verification they are almost always modified and transformed, so as to bring them into conformity with the facts. And not unfrequently these transformations are different in different departments of scientific investigation, in each of which the hypotheses are tested by different methods, and confronted with different orders of facts. The result is, that in many cases not only serious discrepancies but radical inconsistencies are developed between the several forms into which the hypotheses are forced on different lines of research. A very good illustration of this is afforded by the hypothetical æther, which has played a part more or less conspicuous in physical astronomy, in ordinary physics, and in chemistry. By the astronomers, this æther was originally regarded as a fluid of extreme tenuity and mobility, offering no sensible resistance to the movements of celestial bodies; and the question of its continuity or discontinuity was not seriously mooted. Its main function in modern astronomy has been to serve as a basis for hydrodynamical theories of gravitation. In physics, this fluid for some time appeared in several *rôles* in connection with the "imponderables," some physicists going so far as to identify it with one or more of them. But since the promulgation of the kinetic theories of these imponderables, and especially of the dynamical theories of heat, it has been in requisition chiefly in optics as a substratum for luminar undulations. And here, to account for the dispersion of light, physicists came to insist upon its atomic or molecular structure, finding it necessary to assume that the particles of the æther were separated by finite intervals bearing a sensible ratio to the length of a luminar wave. Moreover, they had to endow it with an enormous elasticity, so that its resistance to deformation far exceeded that of the most rigid-elastic bodies. But presently, in other departments of physics, the admission of the molecular or atomic constitution of the æther led to consequences subversive of a number of well-ascertained facts, to some of which I have referred in my seventh chapter. And in chemistry, too, it was found impossible to concede the enormous elasticity of the æther without depriving it of those properties upon which its serviceableness in the construction of chemical theories mainly depended. Furthermore, the exigencies of the atomo-mechanical theory have led distinguished mathematicians and physicists to attempt a substitution, for the traditional atoms of matter, of peculiar forms of vortical motion in a universal, homogeneous, incompressible, and *continuous* material medium, which (unless the attribute of impenetrability is to be dismissed from the concept of matter) must, of course, be identical with the all-pervading æther.

Now, when two hypotheses are radically inconsistent with each other, one or both of them must eventually be discarded. It is true that in certain cases hypotheses which have proved to be untenable, may continue to be serviceable as "working hypotheses," in a secondary sense, as mere devices for holding together facts which have been collected by means of or with reference to them. So long as they are employed for this purpose alone, and with the clear understanding that they are not propounded for any other, there can be no serious objection to their use. But it is otherwise when the specialist seeks to obtrude his own particular hypothetical figment as a finality upon science generally, and to make it the basis of assertions respecting the ultimate constitution of things, and the universal order of nature. It must not be forgotten that the several departments of science are simply arbitrary divisions of science at large, and that their extent and limits are representative of nothing more than the necessities and conveniences of the division of labor. In these several departments the same physical object may be considered under different aspects. The physicist may study its molecular relations, while the chemist determines its atomic constitution. But when they both deal with the same element or agent, it can not have one set of properties in physics, and another set contradictory of them in chemistry. If the physicist and chemist alike assume the existence of ultimate atoms absolutely invariable in bulk and weight, the atom can not be a cube or oblate spheroid for physical, and a sphere for chemical purposes. And a group of constant atoms can not be an aggregate of extended and absolutely inert and impenetrable masses in a crucible or retort, and a system of mere centers of force as part of a magnet or of a Clamond battery. The universal æther can not be soft and mobile to please the chemist, and rigid-elastic to satisfy the physicist; it can not be continuous at the command of Sir William Thomson, and discontinuous on the suggestion of Cauchy or Fresnel.*

I am told that scientific men are fully aware of the provisional and tentative character of their theories, and

* Since writing the above I have met with an admirable passage relating to the same subject in an article contributed by the eminent physicist, *G. A. Hirn*, to the forty-third volume of the *Mémoires de l'Académie Royale de Belgique*, in which he says:

"Lorsqu'on voit l'assurance avec laquelle s'affirment aujourd'hui les doctrines qui rapportent aux seuls mouvements de l'atome l'ensemble, l'universalité des phénomènes, on est en droit de s'attendre à ce qu'il y ait unanimité aussi sur les qualités qu'on assigne à cet être unique, fondement de toute existence. Or dès le premier examen des systèmes particuliers proposés, on éprouve la plus étrange déception; on s'aperçoit que l'atome du chimiste, celui du physicien, celui du métaphysicien, celui du mathématicien . . . n'ont absolument de commun que le nom! Résultat inévitable du morcellement actuel de nos sciences, chacun, dans son casier, se fabrique un atome qui satisfait aux exigences des phénomènes qu'il étudie, sans s'inquiéter le moins du monde des exigences propres aux phénomènes du casier voisin. Le métaphysicien bannit les principes de l'attraction, de la répulsion, comme des rêves: le mathématicien, qui analyse les lois de l'élasticité, celles de la propagation de la lumière, les admet implicitement, sans même les nommer (jamais on n'expliquera la propagation longitudinale des vibrations transversales admises en optique, sans rendre les atomes vibrants solidaires par quelque chose d'autre que les atomes eux-mêmes). Le chimiste ne peut expliquer le groupement des atomes dans ses molécules souvent si compliquées, sans attribuer à ses atomes des qualités spécifiques qui les distinguent: *pour le physicien et le métaphysicien, partisans des doctrines modernes, l'atome est, au contraire, toujours et partout le même.* Que dis-je! on n'est pas plus d'accord dans une seule et même science sur les qualités de l'atome. Chacun le construit à sa guise pour l'explication de tel phénomène restreint dont il s'occupe en particulier."—(*Recherches expérimentales sur la relation qui existe entre la résistance de l'air et sa température, p. 68.*)

carefully refrain from making assertions respecting their ultimate validity. This is, no doubt, true as to a great number of minor hypotheses; but it is far from being true as to those theories and concepts which are the subject of my discussion. If any proof of this were necessary, it would be afforded by the violence with which my "assaults" upon the atomic theory and the kinetic theory of gases, and upon the current misinterpretation of the principle of inertia which lies at the base of the constantly recurring phrase of *dead matter*, etc., have been repelled. So thoroughly axiomatic have the doctrines of the absolutely independent and passive existence of matter, and of the constitution of bodies as aggregates of absolutely constant physical units, become in the minds of modern physicists, that they not only regard them as the indispensable foundations of the whole structure of physical science, but do not hesitate to use them as supports for professorial chairs of metaphysics, from which they promulgate doctrines like those set forth in the "Unseen Universe," or of pulpits from which they preach dogmatic theology. And no thoughtful reader of Du Bois-Reymond's "*Grenzen des Naturerkennens*," or his "*Sieben Weltraethsel*," can be in any doubt as to the degree of certitude ascribed by the highest scientific authorities to the fundamental principles of the atomo-mechanical theory. There are, of course, men of science who do not share this confidence in the absolute and final truth of the theories in question. But the great majority of them deem it a sufficient reply to all suggestions respecting their relative or provisional value to point with pride to the enormous success which has attended physical research since their adoption. In view of this it may be well to examine for a moment to what this success is really due.

All knowledge results from the establishment of relations between phenomena. And systematic knowledge, *science*, results from the establishment of definite relations between phenomena distinctly and clearly apprehended. One of the great vices of the old metaphysical mode of reasoning was, that it operated with vague and abstract concepts, without direct reference to distinct phenomena. Hence the conclusions of metaphysics were as confused, indefinite, and fallacious as its premisses. In contrast to this, the empirical method, proceeding from the observation of particular facts, and basing its cautious and slow progress upon the identification of new phenomena with those previously and familiarly known, avoids this vagueness and confusion. It is in constant sight, and under perpetual control of facts, which act as ever-present checks upon the vagaries of speculation. In a certain sense, the empirical method of science conforms to the natural methods of ordinary thought and speech. It is a well-known saying of Aristotle that all thought is a succession of pictures; and it is a theorem of modern linguistics that forms of speech are a succession of metaphors. The difference between the concepts and theories embodied in the words of a language, and the concepts and theories of science, lies in this, that the former are grounded upon superficial, partial, and often fanciful analogies; whereas the latter are based upon wide generalizations, or upon classifications according to resemblances and correspondences evidencing essential and truly significant relations, such as those of origin, structure, functional equivalence, and the like. Now, both the formation of concepts and words in the evolution of ordinary thought and speech, and the construction of physical theories, are liable to the great structural fallacy which

I have pointed out in the eleventh chapter, and which is due to the fact that, by comparing and partially or wholly identifying phenomena that are new and strange with those that are old and familiar, the first crude and superficial apprehensions of fact are made both the starting-points and the goals of scientific explanation. It is this fallacy which lies at the bottom of the demand of Professor Lodge and Mr. MacAlister (to which I shall have to refer hereafter), "to see the traces" by which the sun's pull upon the planets is exerted. Scientific theorists, no less than ordinary men, incessantly lose sight of the circumstance that, while it is true that the advance of knowledge depends upon the progressive identification of phenomena, so that the most comprehensive theory, the most exhaustive analysis, and the profoundest insight can never yield anything more than ultimate facts—in which sense it has been well said that all science is a classification of mysteries;—nevertheless, it does not follow that the ultimate fact, which lies at the point of convergence of the several lines of identification, is necessarily the fact first observed in the infancy of human intelligence; but that, on the contrary, the highest generalizations and the profoundest theories lead to typical facts which are generally so obscure and difficult of apprehension, that they are brought to light only on the later stages of intellectual development. But this very fallacy attending the habitual explanation of unfamiliar facts, by reference to, and identification with, old and familiar facts, is to a certain extent inseparable from the practical serviceableness of ordinary physical theories. It is the indispensable prerequisite of that incessant play of the "scientific imagination," upon which Professor Tyndall discourses with so much eloquence. The scientific imagination, no less than the ordinary phantasy, is restricted to the elements furnished by sensible experience, which it may group and combine, but can not create. Its play, therefore, is vivid, and, as an aid to the prosecution of physical research (which depends largely upon the pursuit of the analogies of experience), effective in proportion to our familiarity with these elements. This very circumstance, however, in most instances, seriously impairs the ultimate value of the theories founded upon it. The vivid play of the "scientific imagination" depends upon the identification of phenomena with familiar facts; the truth of a theory, on the contrary, upon their identification with recondite, unfamiliar facts, so that the "sensuous picturability," as it has been called, of a physical explanation is generally in inverse ratio to its truth, and the success which has attended the use of a particular hypothesis in the progress of a special scientific inquiry, is by no means a proper measure of its real scientific value.

I have already said that the complaints of my critics of my failure to do justice to "working hypotheses" are not only irrelevant, but also unfounded. Indeed, I am wholly at a loss to see what justification there is in the pages of my book for the charge of the reviewer in the New York "Nation," that I fail to comprehend that physical theories are "mere formal, explanatory, didactic devices"; that "atomism is a symbolical, and later a graphic, system, which might almost be visualized illustratively for specific purposes" (so that, as it would seem, there are, in the opinion of our critic, graphic systems that can not be "illustratively visualized"), etc., etc. In the seventh and eighth chapters of this volume (pp. 85 and 105–116) the reader will meet with a somewhat careful exposition of the nature and function of a scientific hypothesis; while the concluding chapter (pp. 295–299) contains an elaborate discussion of the necessity and value of provisional explanatory devices. And on page 101 he will find this language respecting the atomic theory:

"The foregoing considerations do not, of course, detract from the merits of the atomic hypothesis as a graphic or expository device—as an aid to the representative faculty in 'realizing' the phases of chemical or physical transformation. It is a fact beyond dispute that chemistry owes a great part of its practical advance to its use, and that the structural formulæ founded upon it have enabled the chemist, not merely to trace the connection and mutual dependence of the various stages in the metamorphosis of 'elements' and 'compounds' so called, but in many cases (such as that of the hydrocarbon series in organic chemistry) successfully to anticipate the results of experimental research."

I hope that no presumption, either of ignorance or of a want of literary taste, arises from the fact that I failed to anticipate the peculiar phraseology in which my critic sees fit to paraphrase my own words, when he speaks of the "flexile and serviceable adaptability [of a theory] to many and diverse orders," notwithstanding their want of "exiguous consistency," and of the "radically different tests of canonicity" (whatever that may mean) to be applied to "a material, a working, and an ultimate hypothesis." *

* As exemplifying the humors of modern criticism, it may be mentioned that my critic in "The Nation," in animadverting upon my treatment of the doctrines of pangeometry, has this passage:

". . . *Although he cites Stumpf*, he has not grasped the psychological problem involved in a geometry without parallels."

From this the reader will, no doubt, infer that my critic had read

II.

I NOW come to the consideration of the objections which have been urged against the premisses and conclusions of my general argument. The persons by whom these objections are made may be distributed into two classes: Those belonging to the first class admit that the atomo-mechanical theory is generally held to be the basis of modern physics; but they deny, either that the four propositions which I have enumerated as necessarily involved in that theory—notably the first, second, and fourth—are essential parts or consequences of that theory, or that the facts established by modern physical research, and the generalizations based upon these facts, are in conflict with them. The objectors of the second class, on the contrary, dispute the general thesis that modern physics profess to be founded on the theory in question.

In reference to the objectors of both these classes, but especially those of the first, it is to be said, at the outset, that the professed antagonism of science to metaphysical speculation has led the majority of scientific specialists to assume that the methods and results of empirical research are wholly independent of the control of the laws of thought. They either silently ignore, or openly repudiate, the simplest canons of logic, including the laws of non-contradiction and excluded

Stumpf's book, and was familiar with its contents. Now, the fact is, that the book in question treats of a subject wholly foreign to the pangeometrical question, and its author does not make the most distant reference to any of the "problems involved in a geometry without parallels," or to any topic the discussion of which could throw the faintest light on such a "problem." No one would be more astonished than Dr. Stumpf, if he saw the article in "The Nation," and found himself cited as an authority on the subject of a "geometry without parallels."

middle, and resent, with the utmost vehemence, every application of the rule of consistency to their hypotheses and theories. They are apparently without the least suspicion that not only the theoretical evaluation of the data of experience, but experience itself, is impossible, except on the assumption of definite, universal, and inexorable laws of cognition, and that, but for this assumption, the framing of hypotheses and theories is utterly senseless and vain; and they regard an examination of their hypotheses and theories, in the light of these laws, as an impertinent intrusion of "*a priori* principles and methods" into the domains of empirical science. Persons of this cast of mind find no difficulty in holding that atoms are absolutely inert, and, at the same time, asserting that these atoms are perfectly elastic; or in maintaining that the physical universe, in its last analysis, resolves itself into "dead" matter and motion, and yet denying that all physical energy is in reality kinetic; or in proclaiming that all phenomenal differences in the objective world are ultimately due to the various motions of absolutely simple material units, and, nevertheless, repudiating the proposition that these units are equal.

An admirable illustration of all this is afforded by Mr. Donald MacAlister, who, curiously enough, writes the review of my book for *Mind*, a journal which, according to the announcement of its title-page, is a "Quarterly Review of Psychology and Philosophy." Mr. MacAlister denies the first two propositions of the atomo-mechanical theory, i. e., those asserting the equality of the elements of mass, and their rigidity. He asks, "on what grounds" these propositions are advanced "as essential doctrines of modern physics," and points out that the number of scientific men whose writings I quote in support of them is, after all, comparatively insignificant. He evidently thinks that the only ground upon which I assert the propositions in question to be essential parts of the atomo-mechanical theory (not of *modern physics*, as my critic has it) is, that they are broached by certain prominent scientific men. "It would be easy," he says, "to name half a dozen, with Maxwell and Thomson at their head, who theorize as if the exact opposite were the truth." Now, Mr. MacAlister will, no doubt, be very much astonished when I tell him that this accords precisely with what I say, viz., that these propositions are *not* held by the great body of scientific men who theorize in the presence of the facts, for the simple reason that the facts are inconsistent with them. What I maintain is, that the majority of physicists hold a general doctrine which I designate as the atomo-mechanical theory, from which these propositions inevitably follow; but that, when they construct their special theories, either by generalizing the facts of experience, or by framing hypotheses to account for them, they are constrained to discard and repudiate that doctrine. The authorities adduced in support of the several "quadrilateral" propositions were cited simply for the purpose of showing that these propositions are recognized as necessary corollaries of the atomo-mechanical theory by clear-headed men of science who do not share the delusion that science requires, not merely the substitution of empirical research for the old attempts at reaching physical truths by an analysis of ontological concepts, but also the rejection of the canons of logic.

I may say here, incidentally, that the number of citations from the writings of eminent men of science in support of the propositions just referred to, might have been extended. Among those, for instance, who assert the equality of the elements of mass, is Professor Emil Du Bois-Reymond, as appears from the passage I have already quoted in this introduction for another purpose. And in France, not only the well-known explicit declarations of M. Marignac, but also the ceaseless recurrence among French physicists and chemists of the attempt to exhibit the chemical elements as compounds or allotropic forms of some single element, afford striking proof of the logical constraint, resulting from the atomo-mechanical theory, to reduce the primordial atoms to uniformity and equality.*

While Mr. MacAlister denies that modern physicists assent to the first and second propositions of the atomo-mechanical theory, he admits that they insist upon the third and fourth. But here he contests my assertion, that these propositions are in conflict with the teachings of the sciences of physics and astronomy. He concedes, indeed, that thus far no attempt to account for gravitation, without assuming action at a distance, has met with success; but he contends that there is promise of such success in the vortex-atom theory of Sir William Thomson, my criticism of which he deems wholly inadequate and inconclusive, because it is, as he asserts, ontological and based on *a priori* considerations. Mr. MacAlister's treatment of this subject is so characteristic of the confusion of modern theorists, who insist upon reducing all physical action to impact, that it is, perhaps, not improper to examine it at some length.

* One of the curious animadversions of Mr. MacAlister is, that I quote some of the declarations of scientific men respecting the equality of the elements of mass from magazine articles. I am happy to inform him that he may now read the opinions of Professor Wundt in a bulky volume—vol. ii of his "Logik"—which has just left the press.

Mr. MacAlister enforces and illustrates his claim, according to which no theory of gravitation is valid which does not account for it upon the principles of impact or pressure, by the following quotation from a lecture of Professor Lodge:*

"If a man explained the action of a horse on a cart by saying that there was an attraction between them, varying as some high direct power of the distance, he would not be saying other than the truth—the facts may be so expressed; but he would be felt to be giving a wretchedly lame explanation, and any one who simply pointed out the traces, would be going much more to the root of the matter. Similarly with the attraction of a magnet for another magnetic pole. To say that there is an attraction as the inverse cube of the distance between them, is true, but it is not the whole truth; and we should be obliged to any one who will point out the traces, for traces we feel sure there are."

The passage here adduced is followed, in Professor Lodge's lecture, by others which Mr. MacAlister does not quote. The lecturer proceeds to say, among other things, that—

"A pull resolves itself into a push; to pull a thing toward you, you have to put your finger behind it, and push; a horse is said to pull a cart, but he is really pushing at the collar; an engine pushes a truck by means of a hook and eye, and so on. *There is still the further very important question as to why the parts hang together, and why, when you push one part, the rest follows. Cohesion is a very striking fact, and an explanation of it is much to be desired. I shall have a little more to say about it later; at present we have nothing more than an indication of the direction in which an explanation seems possible.*"

In another place in the same lecture:

"Metaphysical arguments, in so far as they have any weight or validity whatever, are unconscious appeals to experience; a person endeavors to find out whether a certain condition of things

* "Nature," vol. xxvii, p. 304.

is by him conceivable, and, if it is not conceivable, he has some *prima facie* ground for asserting that it probably does not exist. . . . If a highly-developed mind, or set of minds, find a doctrine about some comparatively simple and fundamental matter absolutely unthinkable, it is an evidence, and is accepted as good evidence, that the unthinkable state of things has no existence; the argument being that, if it did exist, either it, or something not wholly unlike it, would have come within the range of experience. We have no further evidence than this for the statement that two straight lines can not inclose a space, or that three angles of a triangle are equal to two right angles."

And thereupon, toward the end of his lecture, Professor Lodge indicates that the explanation of cohesion, as well as of gravity, is to be looked for in the vortex-atom theory of Sir William Thomson.

Now, what is the gist of all this reasoning? Why is it necessary, in the opinion of Professor Lodge and Mr. MacAlister, to "point out the traces" in accounting for gravitation? Obviously for no other reason than this, that every true account of a physical phenomenon is, in its nature, an exhibition of its identity with some familiar fact of experience. In the language of Newton, the cause to which it is referred must be a *vera causa*. *Actio in distans*, according to the claim of Professor Lodge and Mr. MacAlister, is not a fact of familiar experience—indeed, it is not, in any proper sense, a fact at all—while a pull, by means of a continuous line, the parts of which cohere, or rather a push, by means of a continuous rod, is such a fact.

In discussing this, I will not stop to inquire whether it be true or not that distant action is not a familiar fact; or, indeed, whether we have any experience of physical action which, on close examination, does not resolve itself into *actio in distans*. I will content myself with inquiring whether or not the elements of the vortex-atom theory are familiar, or even possible, facts of experience. For, if they are not, clearly that theory is obnoxious to the same criticism which is said to invalidate the assumption of *actio in distans.*

The medium in which the vortex-movements arise is, according to Professor Lodge's own express statement,* "a perfectly homogeneous, incompressible, continuous body, incapable of being resolved into simple elements or atoms; it is, in fact, continuous, not molecular." And, after making this statement, Professor Lodge adds, "*There is no other body of which we can say this, and hence the properties of the æther must be somewhat different from those of ordinary matter.*" It appears, then, that the whole vortex-atom theory, which is offered to us as a substitute for the "metaphysical theory" of *actio in distans,* rests upon the hypothesis of the existence of a material medium which is utterly unknown to experience, and which has properties *somewhat* † different from those of ordinary matter. Hence this theory, instead of being, as is claimed, a reduction of an unfamiliar fact of experience to a familiar fact, is, on the contrary, a reduction of a fact which is perfectly familiar, to a fact which is not only unfamiliar, but wholly unknown, unobserved, and unobservable.

Furthermore, the alleged vortical motion of, or rather in, the assumed æthereal medium is, as I have shown on pages 43 and 44 of this volume, *impossible,* because

* "Nature," vol. xxvii, p. 305.

† *Somewhat* different! The real import of this "somewhat" is, that the medium in question is not, in any intelligible sense, material at all, having none of the properties of matter. All the properties of matter depend upon differences and changes, and the hypothetical æther here defined is not only destitute of differences, but incapable of difference and change.

"motion in a perfectly homogeneous, incompressible, and therefore continuous fluid, is not sensible motion." Mr. MacAlister again attempts to break the force of this simple and obvious reflection, by stigmatizing it as "ontological," and (in his ignorance of the meanings of ordinary logical and psychological terms, construing the word *sensible* as denoting simply what can be *felt*) admonishes me that, on proper occasion, "the impact of a vortex-atom stick on a vortex-atom head may quite well give rise to the idea of 'sensible motion' in the head." But, if he will refer once more to Professor Lodge's lecture, he will find that his authority, in one of the passages just quoted, distinctly sets up the criterion of conceivability as a test of probable or possible existence; and even contends that by this test alone can we establish the truth of a geometrical proposition. What is not thinkable can not, according to the doctrine of Professor Lodge, be true. In the light of this doctrine let us examine, then, whether or not motion in an absolutely homogeneous, continuous, and incompressible medium, is thinkable.

Observe: it is not some thing or body distinct from the medium whose motion is to be conceived; it is motion of and in the medium itself. Now, the conception of real (i. e., sensible) motion of necessity involves three distinct elements, viz.: *a moving thing or body; a place from which it moves; and a place to which it moves.* And these elements or data are prerequisites to the conception, and, therefore (according to Professor Lodge's own criteria), to the possible reality of motion. They must not only be discriminated by some mark or attribute from the medium in which the motion occurs, but they must be so discriminated before the occurrence of the motion, and independently of it; their identification can not be the result of the motion itself. But all this is expressly negatived or precluded by the hypothesis of the absolute homogeneity, continuity, and incompressibility of the medium, and of the total absence of any thing or physical entity which could serve as the identifiable substratum of motion, or as an object or point of reference. It is manifest, therefore, upon the considerations presented by my critic and Professor Lodge themselves that, wherever the vortex-atom theory may land us, it certainly does not land us anywhere in the region of physics, or in the domain of *veræ causæ*. And I may add that, inasmuch as the hypothetical undifferentiated and undifferentiable medium is clearly an involuntary reification of the old ontological concept *pure being*, the theory under discussion has all the attributes of an inapprehensible metaphysical phantom.

It ought to be noticed, perhaps, in passing, that the argument just presented, from my critic's own premisses, against the possibility of actual motion in the hypothetical medium, is a valid argument, although one of these premisses—the doctrine of Professor Lodge that inconceivability, or "unthinkableness," of a particular state of things by a "highly-developed" mind, or group of minds, is proof of its impossibility—is unsound. As I have shown, in the ninth chapter, inconceivability is proof of impossibility only in case the concept attempted to be formed requires the union of contradictory attributes. The attempt to conceive motion as taking place in the medium referred to obviously presents that case: in asserting that its parts are capable of real motion, it impliedly invests the medium with attributes of which it is expressly deprived by the negations by which alone it is defined.

Having disposed of the objections of my critic in *Mind*, so far as, in my judgment, they are deserving of notice, I proceed to consider another objection which has been urged by certain specialists: the denial of the necessity of attributing elasticity to the ultimate atoms of matter. It is said that all that is required to be assigned to the ultimate atoms is, "an insuperable repulsive force." The first obvious reflection respecting this claim is, that an indefinite, insuperable force, if it be a real force of repulsion, and not merely a passive resistance to intrusion into the space occupied by the atom (i. e., impenetrability), is inconsistent with the conservation of energy. If, on the other hand, it is a definite mechanical function, it is clear that such a force is but another name for elasticity, and that the proposed substitution of a new name for the old one does not, in the least degree, affect the validity of my argument. There are physicists, however, who imagine that the fact of resilience in cases of atomic impact is compatible with the theory of the absolute rigidity of the atoms; and, in support of this proposition, they invoke the authority of Dr. G. Luebeck * and O. E. Meyer.† But, on examination of Dr. Luebeck's article, it appears that he recognizes the necessity of attributing elasticity to the atoms, if they are regarded as *bodies*, and seeks to avoid this necessity by contending that they are not bodies, asserting that, in view of this fact, it follows from the formulæ respecting the impact of both elastic and inelastic bodies, *in conjunction with the principle of the conservaton of vis viva*, that the impact of atoms must result in resilience. Stated in simple words, his argu-

* "Ableitung des elastischen Stosses zweier Atome aus mechanischen Principien," Schloemilch's Zeitschrift f. M. u. P., xxii, 126.

† Die kinetische Theorie der Gase, third edition, p. 237, *seq.*

ment is that, inasmuch as in the case of the impact of ordinary rigid bodies consisting of parts, the translatory motion of these bodies may be converted into molecular motion, and inasmuch as such conversion is impossible in the case of simple atoms, therefore it is necessary, under the coercion of the principle of the conservation of *vis viva*, to frame a concept of the atom which includes the attribute of resilience. It is hardly necessary to observe that this is not a refutation but an enforcement of the doctrine according to which the ultimate material constituents of a body must have the property of elasticity, because without it their collision would be destructive of energy—that the absolute inertia and rigidity of these constituents is inconsistent with the conservation of energy. This evidently has not wholly escaped the apprehension of O. E. Meyer; for, though he refers to Dr. Luebeck's article, he observes that, in his opinion, the simplest and most probable supposition is that, "*at the moment of the impact of absolutely rigid bodies, their rigidity and the impossibility of compressing them suddenly give rise to a repellent force.*" * Of this strange theory it is sufficient to remark that, if projectile forces can suddenly and spontaneously arise upon the mere contact of hard bodies from the impossibility of compressing them, we have sources of physical energy which will very materially extend the catalogue of Sir William Thomson.

The last objection which I propose to consider is very strenuously urged, not only by opponents but

* . . . bei dem Zusammenstoss absolut harter Koerper, aus deren Haerte und der Unmoeglichkeit sie zusammenzudruecken, im Momente des Stosses ploetzlich eine zurueckstossende Kraft entsteht. Die letztere Annahme halte ich fuer die einfachste und wahrscheinlichste. *Loc. cit.*, p. 239.

also by adherents of the atomo-mechanical theory, and consists in the denial of my statement that, in the light of that theory, all potential energy is in reality kinetic. Those who urge this objection do not attempt to show that there is any escape from this proposition, if all the phenomena of the world are ultimately resolvable into matter and motion, and if matter is absolutely inert, *dead*. Nevertheless, they say that the proposition in question is the ill-considered and hastily-formed opinion of Professor P. G. Tait, for which physicists generally are not to be held responsible.

Now, in the first place, the opinion of Professor Tait is neither ill-considered nor hastily formed. That it is his deliberate opinion appears sufficiently from the fact that he has very recently repeated and enforced it by new considerations in the article on "Mechanics" contributed by him to the ninth edition of the "Encyclopædia Britannica." And, in the second place, he is by no means alone in that opinion. One of the most noted physicists and chemists in the United States is Professor G. F. Barker, of the University of Pennsylvania, who, in the address delivered by him as retiring President of the American Association for the Advancement of Science, in 1880, said: *

"As defined hitherto, energy is either motion or position; is kinetic or potential. Energy of position derives its value obviously from the fact that, in virtue of attraction, it may become energy of motion. But attraction implies action at a distance; and action at a distance implies that matter may act where it is not. This, of course, is impossible; and hence action at a distance, and with it attraction and potential energy, are disappearing from the language of science. . . . Now, as Preston has suggested, if we regard the æther as a gas, defined by the kinetic theory that its molecules move in straight lines, but with an

* "Popular Science Monthly," October, 1880, pp. 766, 767.

enormous length of free path, it is obvious that this æther may be clearly conceived of as the source of all ordinary matter. It is an enormous store-house of energy, which is continually passing to and from ordinary matter, precisely as we know it to do in the case of radiant transmission. *Before so simple a conception as this, both potential energy and action at a distance are easily given up. All energy is kinetic energy, the energy of motion.*"

And in England we have Herbert Spencer, who, though not a professional physicist, is regarded by a great number of eminent men of science as the highest authority on matters like that now under discussion. While he is generally at war with Professor Tait, he is in full agreement with him here. In the "Appendix" to the fourth edition of his "First Principles,"* Mr. Spencer, replying to certain criticisms of Professor Birks, says:

"Now, the tacit implication here is, that I accept the doctrine of potential energy. . . . In the first place, I have to ask on what authority Professor Birks assumes that I hold the doctrine of potential energy in the way in which it is held by those named? . . . In the chapter on 'The Continuity of Motion,' I have, at considerable length, given reasons for regarding the conception of potential energy as an illegitimate one; and have distinctly stated that I am at issue with scientific friends on the matter. . . . Let me add that my rejection of this doctrine is not without other warrant than my own. Since the issue of the last edition of this work, . . . Mr. James Croll, no mean authority as a mathematician and a physicist, has published, in the 'Philosophical Magazine' for October, 1876, page 241, a paper in which he shows, I think conclusively, that the commonly accepted view of potential energy can not be sustained, but that energy invariably remains actual."

Although, as I have said, the denial of the essentially kinetic character of physical energy proceeds, not only from those who reject the atomo-mechanical

* Pp. 583, 584.

theory, but likewise from many who recognize it as the true basis of modern physics, yet this denial, on the whole, is the point of divergence between two distinct schools of physicists. With few exceptions, the adherents of both schools agree in regarding the ultimate constituents of the material world as absolutely indivisible, inert atoms of invariable mass and volume, whose constancy in number (in conjunction with that of mass and volume) constitutes the conservation or indestructibility of matter. In order to account for the differences and changes in the material world, this inert element, *matter*, obviously requires to be supplemented by another element; and this element, according to the adherents of the atomo-mechanical theory, is *motion*, whereas, in the view of its opponents, it is *force*. This force is assumed to be an independent, substantial entity, which is not a property of matter, nor essentially related to matter, otherwise than by its power to act upon it. Force, it is said, resides, not in the atoms, but in the space between them. The most distinguished representative of this school is G. A. Hirn, the well-known author of the *Théorie Mécanique de la Chaleur*, in whose writings the independent substantiality of force is stated and illustrated in a variety of ways, of which the following may serve as an example: *

"The question," says M. Hirn, "brought to its simplest expression, reduces itself to the inquiry, Whether force resides only in the atom, or outside of it? . . . Is force in the material atom, or in the space which separates two atoms? . . . By the aid of the data of the mechanical theory of heat I have demonstrated that matter can not be regarded as infinitely divisible; that the atom of the chemists is not an entity of pure convention, and simply

* *Conséquences philosophiques et métaphysiques de la Thermodynamique*, p. 61, *seq.*

an explicative device, but that it exists really, that its volume is absolutely unalterable, and that, consequently, it is not elastic. Force, therefore, is not *in* the atom; it is *in the space* which separates the atoms from each other."

In the light of this doctrine gravitation, cohesion, affinity, etc., are, of course, not due to any real *actio in distans* exerted by the atoms upon each other, but to the forces which continuously fill the spaces between them. To what remarkable consequences this theory leads, may be seen from the article of Dr. James Croll,* already referred to, in the quotation from the Appendix to the "First Principles" of Herbert Spencer, in which it is said:

"The fact that gravity increases inversely as the square of the distance may be regarded as evidence of the truth of the views advocated by Faraday, Waterston, and others, that it is a force pervading space external to bodies, and that on the mutual approach of the bodies the force is not increased, as is generally supposed, but the bodies merely pass into a place where the force exists with greater intensity. . . . When a stone, projected upward, recedes from the earth, its *vis viva* is transferred to space, and exists there as gravity. When the stone approaches the earth, the force existing in space is transferred back to the body, and reappears as *vis viva*."

I cite these passages for the purpose, not of entering upon a lengthy discussion of the doctrine set forth in them, but simply of indicating its character. There are two suggestions respecting it, however, which I desire to make in passing, without dwelling upon them. The first is, that the substantialization of force here presented rests upon an oblivion of the fact that all force is essentially a stress—an action between two bodies. The other suggestion is that, if the increase of

* Croll, *On the Transformation of Gravity*, *Phil. Mag.* [*V*], vol. ii, p. 252.

the velocity of falling bodies is due to the circumstance that they pass into places "where the force exists with greater intensity"—if bodies projected against gravity transfer their *vis viva* to space where it is left as *force*, and take it up again, reconverting it into *vis viva* as they return to the same points, it follows that each point in space has an infinite store of forces of all degrees of intensity, so as to supply the countless bodies gravitating toward each other in all directions and from all distances, each with the amount of gravitative force due to it at that point.*

It is proper to observe here that, while the substantialization of force is, in my judgment, an ontological error, nevertheless the objections urged by M. Hirn and others to the assertion that the phenomena of the universe can be reduced to the motions of inert atoms, appear to me worthy of the most serious attention. M. Hirn has recently, since the appearance of the first edi-

* The blindness of eminent physicists to some of the most obvious consequences of their own theories is marvelous. On page 54 of this volume I have cited the twenty-first query appended by Newton to his "Opticks," in which he suggests an explanation of gravity on the supposition that the attracting and attracted bodies are surrounded by æthereal media increasing in density from the centers of such bodies outward. Now, nothing seems to be more manifest than that this theory requires a separate and distinct medium for each body, and that a single medium surrounding all bodies can not possibly be so constituted as to increase in density from the center of each body outward. A similar observation applies to many of the assumptions respecting the constitution of the æther, which have been made to account for the phenomena of cohesion, chemical affinity, etc., including those of M. Cauchy in his *Mémoire sur la dispersion de la lumière.* Mr. Walter R. Browne, in a very thoughtful essay on *Action at a Distance* (*Phil. Mag.*, December, 1880), has pointed out that the several kinetic theories of gravitation, cohesion, magnetism, etc., require a different and distinct "gravity-gas," "cohesion-gas," "magnetism-gas," etc., for each of the phenomena sought to be explained by the theory of impact.

tion of this volume, published an article on the results of a series of experiments made by him to determine the relation between the resistance of the air and its temperature,* in which he arrives at the conclusion, *on experimental grounds*, that the kinetic theory of gases, and "all explicative hypotheses which pretend to account for the phenomena of heat by the motions of ponderable atoms," are untenable and must be abandoned. He shows, by an analysis and development of the formulæ of Clausius himself, that, according to the kinetic theory of gases, the resistance of air (or any other gas) to the motion of a body is not merely a *mediate* or indirect function of its temperature (i. e., by reason of the effect of variations of temperature upon the density of the gas), but that it is an *immediate* or direct function of the temperature—otherwise expressed, that in gases of constant density the resistance necessarily varies with the temperature, if the kinetic theory be true; whereas, on the contrary, it is the unquestionable result of four distinct lines of experiment that, in fact, the resistance does *not* vary with the temperature so long as the density remains the same. To leave no doubt as to the comprehensiveness of his proposition that "the pressure and temperature of a gas do not consist in the motions of material atoms," M. Hirn adds: "I say *whatever may be the nature of the motion.* For the rectilinear and translatory motions postulated in the particular theory we have discussed [the kinetic theory of Clausius], we may substitute any other imaginable motion: vibratory motion, gyratory motion (molecular vortices), . . . the moment these motions simulate a repulsive force they modify in the same way the law of resistance in gases and vapors."

* *Mémoires de l'Académie Royale de Belgique*, tom. xliii.

III.

A CRITICAL observation has been made by persons who correctly apprehend my purpose, and are more or less in sympathy with it, that should not, perhaps, be passed without notice. The observation is, that the nature of my inquiry is, after all, metaphysical, and that I am wrong in identifying metaphysics with ontology.

The question thus raised relates exclusively to the proper use of terms. It is not denied that, up to a very recent period, the speculations commonly designated as metaphysical have proceeded upon ontological assumptions. But it is asserted that there is a "new metaphysic" which has discarded these assumptions, and which, though abandoning the old lines of purely speculative reasoning, and resorting to methods of investigation that have been accredited by their results in special departments of science, nevertheless aims at a solution of many, if not all, of the old problems of metaphysics, or at least at a determination of the limits within which a solution is possible.

I can not but think that this attempt to retain an old name for an essentially new thing is unfortunate, because it leads to intolerable confusion. A very signal illustration of this is afforded by the fate of Kant. It was the distinct and avowed purpose of the *Critique of Pure Reason* to demonstrate the utter futility of metaphysical speculation as it existed in Kant's time; and yet, by reason of his retention of the term "metaphysics" to denote the investigation of the laws which, in his view, were not results but conditions of experience, he is constantly (and not wholly without justice) assigned to the ranks of the ontological metaphysicians.*

But, while I am at issue on this point with the modern vindicators of the name and authority of metaphysics, I am in entire accord with them when they maintain that physical science can not be permanently constructed from the crude results of experiment by the mere aid of mathematical operations. There is a current saying that every problem in physics abuts upon a problem in metaphysics. The real import of this saying, I take it, is that the ultimate interpretation of the data of experiment and observation involves questions which can not be answered by the balance or retort, or

* My antagonism to metaphysics has produced an impression in the minds of some of my readers that I am at bottom an adherent of the "positivistic" philosophy of Comte; and passages are cited from Comte's writings which are analogous to certain propositions in my ninth chapter. There is no doubt that Comte has the great merit of having brought some of the fallacies of metaphysical reasoning into clear relief; and, if it be true that he was unacquainted with the writings of Kant, some of his statements are very remarkable. It has sometimes appeared to me that his English critics, in protesting against the imputation of Comtism, have failed to do him full justice in this respect. But it is proper to say that I am not, so far as I am aware, indebted to Comte for any of the elements of my discussion, and that the propositions which are in substantial accord with his writings are simply part of the modern theory of Cognition, whose fundamental principles are so familiar to those who are acquainted with that theory, that no one, at this day, thinks of citing authority for them. The difficulty with the French adherents of Comte is that, with few exceptions, they are strangers to the logical and epistemological discussions that have been carried on in Germany since the days of Kant and Fries, and therefore credit their master with discoveries which, if he made them independently, were *enfoncements de portes ouvertes*. As to those doctrines whereof Comte is the original promulgator, and the body of which constitutes what is generally known as positivism, it is hardly necessary to inform the intelligent reader that I dissent from them *in toto*. I may add that I have not looked into any of Comte's writings for more than twenty years.

by a mathematical formula, and that this interpretation is possible only in the light of a general theory of cognition. The immediate results of experiment can not become elements of any sort of reasoning, not excepting mathematical reasoning, without being transformed into concepts. And the scope of mathematical reasoning is limited to the determination of the quantitative relations between phenomena or the concepts representing them. The declaration of Professor Tait * that "physical science is to be based entirely upon experiment and mathematical deductions from experiment," ignores the nature and limits, both of experimental inquiry and of mathematical operations. A satisfactory and exhaustive discussion of this topic is impossible here, and must be reserved for a future occasion; but I may, to some extent, anticipate this discussion by indicating some of its elements.

Physical experiments, as well as mathematical deductions from them, are based on certain assumptions respecting the nature of space and the universality of the laws of causality, constancy, and continuity. As to the importance of the assumptions respecting the nature of space, and their bearing upon the interpretation of the data of experiment and observation, it is sufficient to point to the questions hinted at by Professor Tait himself in the passage I have quoted on page 211 of this volume. Without attempting to add anything to what I have said on this subject in the thirteenth and fourteenth chapters, I proceed to a cursory notice of the laws of causality, constancy, and continuity, and of the use made of them in modern physics.

The *law of causality* is essentially a law of the correspondence and equivalence of changes, its root being

* On Some Recent Advances in Physical Science, p. 6.

the essential relativity and mutual dependence of all physical phenomena. It imports that, whenever a change is observed in a given phenomenon or set of phenomena, we are constrained, by reason of the relations of these phenomena to other phenomena upon which their existence depends, to look for an equivalent change in these other phenomena. The question of *cause* never arises, except where there is change; and the cause demanded is likewise a change. This, of course, at once presents the question, What are the criteria of the equivalence required?—a question which can not be examined here, and must be dismissed with the simple remark that these criteria depend upon the nature of the changes whose correspondence and equivalence are under consideration.

If this account of the nature and origin of the law of causality is correct, it is readily seen that the misconceptions of this law by the most distinguished modern physicists are as fatal and prolific of error as those of the ontologists. "The final aim of theoretical physical science," says Helmholtz, "is to find the ultimate unchangeable causes of the processes in nature." * These ultimate unchangeable causes, in the view of the majority of physicists, are simple and unchangeable things or substances, forces, and qualities. "Whenever," says Isenkrahe (after citing the words of Helmholtz just referred to), "natural science succeeds in reducing all changes in the material world to unchangeable qualities of an immutable substance, it has arrived, in my opinion, at the goal of its endeavors." † Similarly Du

* *Das endliche Ziel der theoretischen Naturwissenschaft ist also, die letzten unveraenderlichen Ursachen der Vorgaenge in der Natur aufzusuchen.* Helmholtz, Ueber die Erhaltung der Kraft, p. 2.

† *Kommt die Naturforschung einmal dazu, dass sie alle Veraenderungen*

Bois-Reymond: "Our demand for causality is satisfied only when we conceive matter as being at rest and uniformly distributed in infinite space an infinite time ago." * It is evident that the background of all these statements is the ontological reification of the concept *cause*, analogous to the pangeometrical reification of space.

The *law of constancy* is nothing more than the law of causality in the aspect of the equivalence and reciprocal correspondence of the phenomena determining each other, which, in mechanics, is exemplified in the law of the equality of action and reaction. The necessity of finding a basis for the determination of this equivalence involves the requirement of an ideal or conceptual constant amid phenomenal variation. Here, again, the tendency to ontological reification has led the modern physicist to assert the constancy of absolutely unchangeable finite material elements and of determinate physical forces.

The *law of continuity* is another aspect of the law of causality, resulting from the fact that all physical changes occur in space and time which are necessarily conceived as continuous. To say that space is discontinuous would be to say that between two nearest spatial points there is a spatial interval; and similarly, to say that time is discontinuous would be to say that between two nearest points of time there is an interval of

der Koerperwelt auf unveraenderliche Qualitaeten einer stets sich gleichbleiblenden Substanz zurueckfuehren kann, dann ist sie meiner Meinung nach am Ziel ihres Strebens angelangt. Isenkrahe, Das Raethsel der Schwerkraft, p. 139.

* *Unser Causalitaetsbeduerfniss fuehlt sich nur befriedigt, wenn wir uns vor unendlicher Zeit die Materie ruhend und gleichmaessig im unendlichen Raum vertheilt denken.* Emil Du Bois-Reymond, Die Sieben Welträthsel, p. 10.

time. The vague apprehension of this law, in conjunction with the assumption of the permanent existence of a constant number of discrete atoms of invariable mass and volume, and of a constant sum of forces or motions, has given rise to the standing protest against the possibility of *actio in distans*, and to the various theories respecting the existence of continuous æthereal media, or of forces, between the atoms, in order to account for the continuity of their interaction.

A question of exceeding interest and importance is the old question whether the law of causality, with its several specifications in physical science, including the laws of constancy and continuity, the law of the conservation of energy, the law of least action, the law of motion under least constraint, etc., are purely inductions from experience, or are, on the contrary, conditions precedent to experience, and therefore *in some sense* independent of it. The tendency among modern physicists and mathematicians is decidedly toward the purely empirical view, according to which the only warrant for the assumption of the universality of these laws lies in the fact that thus far no experiential instance has been adduced against them. Helmholtz,* among others, emphasizes this view. But physicists, and especially mathematicians, are puzzled by the circumstance that not only has the law of causality always been applied before any experiential induction was thought of, but that all the other laws above mentioned were announced long before they were precisely formulated and experientially verified, or had been recognized as subjective psychological laws before they were applied to the ob-

* *Fuer die Anwendung des Causalitaetsgesetzes haben wir keine weitere Buergschaft als seinen Erfolg.* Helmholtz, Die Thatsachen in der Wahrnehmung, p. 41.

jective world. This is strikingly illustrated by the law of least action and Gauss's law of motion under least constraint. The law of least action is stated with singular distinctness by Leibnitz;* and Gauss observes in relation to the law of motion under least constraint that "it is remarkable that free motions, when they are inconsistent with the conditions of the system, are modified by Nature in precisely the same way in which the calculating mathematician corrects quantitative results that are necessarily interdependent, by the method of least squares."† It is my belief that the mysteries here presented will be cleared up by a re-examination, upon new grounds, of the nature of the old antithesis between truths *a priori* and cognitions *a posteriori*.

Although it must have become clear, by this time, to those with whom I shall deem it profitable hereafter to have any discussion, that I am not intent on subverting the science of physics or on questioning the facts it has brought to light and the legitimate generalizations based upon them, and, furthermore, that I do not propose to deprive the physicist of his laboratory hypotheses, and hence am under no obligation to supply him with new ones, the question is not unnatural, what concern the physicist has with the discussions in this volume. The simple answer to this is, that physical theories are not merely instrumentalities for the discovery and classification of facts in furtherance of the practical purposes of life, but that they also serve as a basis for the various attempts at a solution of the great questions

* *Semper scilicet est in rebus principium determinationis quod a maximo minimove petendum est, ut nempe maximus præstetur effectus minimo ut sic dicam sumtu.* Leibnitii, Opp. ed. Erdmann, p. 147.

† Ueber ein neues allgemeines Grundgesetz der Mechanik, Gauss' Werke, V, p. 28.

which always have been and always will be the cardinal problems of human thought. And before these questions can be properly submitted to the arbitrament of the physicist, it is necessary that he should have a clear insight into the nature of the concepts and theories with which he operates, and into their relations to the phenomena which they represent. When Du Bois-Reymond, in his *Sieben Weltraethsel*, declares that all questions relating to the "essence" of matter, force, and motion, and the origin of sensation, are necessarily transcendent, while those concerning the nature and origin of life and thought are not, he assumes the absolute objective reality of atoms independently of the intellect; and it certainly is of the greatest importance to ascertain whether, and in what sense, this assumption is true. When the physiological psychologist investigates the laws governing the correspondence between physical and psychological phenomena, he is confronted at the outset with what one of my critics terms the "residual but tedious problem of idealism" with which, in his opinion, I unnecessarily and impertinently weary the reader. The proper statement of that problem, at least, if not its solution, is not "residual," but preliminary to any fruitful discussion of that large number of ultimate questions which are covered by a mass of wordy rubbish to which the physicists have made as large a contribution as the ontological metaphysicians.

Irrespective of this, however, if physics *as a science* are not to fall into utter disrepute, it is time to evoke some order from the confusion which prevails among the very first principles, theories, and definitions of theoretical physics. When Professor Tait, in conjunction with Professor Stewart, announces that "matter is simply passive,"* and then, in connection with Sir William Thomson, declares that "matter has an innate power of resisting external influences,"† it is hardly impertinent to inquire how these statements are to be reconciled. When Professor Du Bois-Reymond, in a passage heretofore quoted, insists upon the necessity of "reducing all the processes of nature to motions of a substantial, indifferent substratum *wholly destitute of quality*,"‡ having declared shortly before in the same lecture that "resolution of all changes in the material world into motions of atoms *caused by their constant central forces* would be the completion of natural science,"# we are in a perplexity from which we have the right to be relieved.

There is, moreover, a lesson to be learned, by the ordinary physicist, from the discussions to which I have invited his attention, as to the limits within which the arbitrary construction of hypotheses and theories, without regard to the validity of the assumptions on which they rest, is useful and proper. This is beginning to be understood by thoughtful physicists, who suggest, for example, that the exhibition of the equivalence of physical and chemical phenomena would lose nothing in clearness if the atomic hypothesis were wholly discarded. An analogous suggestion respecting the basis of statics has recently been made by Professor Horace Lamb,‖ who insists that certain assumptions in the current theories relating to the transmission of force, such as that of the absolute rigidity of the bodies by

* The Unseen Universe, § 104.

† Thomson and Tait, Treatise on Natural Philosophy, vol. i, § 216.

‡ Ueber die Grenzen des Naturerkennens, p. 5.

\# *Ib.*, p. 2.

‖ Phil. Mag. [V], vol. xv, p. 187.

means of which this transmission is effected, are not helps but hindrances to a proper apprehension of the laws of equilibrium and motion, and that the science of statics ought to be constituted by means of the principles of linear and angular momentum, under the definition that "two sets of forces are equivalent when, and only when, they produce the same effect on the linear and angular momentum of any material system to which they are applied." The same spirit is evinced in Kirchhoff's *Lectures on Mathematical Physics*, to which I have already had occasion to refer in another connection. And there is evidence, in the best scientific literature of the day, of a growing tendency to divest the edifice of physical science as far as possible of the hypothetical scaffolding, which not only obstructs the view of its fair proportions, but masks the real principles of its construction, by which its strength and permanence are assured.*

* Before permitting this edition to go to press, it is proper to apologize for an omission in the fifth chapter which I have not deemed it necessary to supply, because it does not affect the validity of my reasoning, and the correction could not be made without serious disturbance of the text. In speaking (pp. 61, 62) of Mr. Adams's revision of the calculations of Laplace respecting the gradual diminution of the excentricity of the earth's orbit due to the disturbing action of the other planets, it ought to have been mentioned that the revision of Mr. Adams has since been revised in turn by Professor John N. Stockwell, of Cleveland, the distinguished author of the *Theory of the Moon's Motion*. I regret this omission the more, because Professor Stockwell is a citizen of my own State.

PREFACE.

The following pages are designed as a contribution, not to physics, nor, certainly, to metaphysics, but to the theory of cognition. Their contents are the result of a somewhat careful study of the true relation of the physical sciences to the general progress of human knowledge. It is the common opinion of contemporary physicists that there was a total breach of continuity in the line of this progress at the point where the thoughts of men were turned from ancient and mediæval traditions respecting the phenomena of nature and their significance to the order and sequence of these phenomena as disclosed by their own observations and experiments, and that the structure of what may, for want of a better name, still be called philosophy now rests upon foundations wholly different from those upon which it stood before the days of Galilei and Bacon. According to this view, Bacon's demand (in the preface to his *Novum Organum*) "that the whole work of the mind be undertaken anew"—*ut opus mentis universum de integro resumatur*—has been thoroughly complied with, and Newton's admonition to the physicists, "to beware of metaphysics," has been effectually heeded. The belief is that modern physical science has not only made its escape from the cloudy regions of metaphysical speculation, and discarded its methods of reasoning, but that it has likewise emancipated itself from the control of its fundamental assumptions. It is my conviction that this belief is but partially conformable to the fact, and that the prevailing misconceptions in regard to the true logical and psychological premisses of science are prolific of errors, whose reaction upon the character and tendencies of modern thought becomes more apparent from day to day. The shallow and sciolistic materialism—I allude, of course, not to its supposed ethical but to its purely intellectual aspects—which for a time threatened to blight the soil and poison the atmosphere even of the old highlands of thought on the continent of Europe, claims to be a presentation of conclusions from the facts and principles established in the several departments of physical science. It is part of my endeavor to meet this claim by an examination of the fundamental concepts and general theories of that department of physical science which is, in a sense, the basis and support of all its other departments —the department of physics. It will be seen at once, upon a most cursory glance at any one of the chapters of this little book, that it is in no wise intended as an open or covert advocacy of a return to metaphysical methods and aims; but that, on the contrary, its tendency is throughout to eliminate from science its latent metaphysical elements, to foster and not to repress the spirit of experimental investigation, and to accredit instead of discrediting the great endeavor of scientific research to gain a sure foothold on solid empirical ground, where the real data of experience may be reduced without ontological prepossessions. An attentive perusal of these pages will make it clear, I think, that this endeavor is continually thwarted by the insidious intrusion into the meditations of the man of science of the old metaphysical spirit. This fact having been established, it was incumbent on me to ascertain, if possible, its causes and, within the narrow limits at my command, to develop its consequences. In the performance of this task it became necessary—inasmuch as I wrote for a class of readers with whom, unfortunately, familiarity with the laws of thought is a somewhat rare accomplishment—to make an excursion into the domain of logic, and to enter upon a brief discussion of the theory of conception. This discussion is, of necessity, very perfunctory, but I venture to hope that it will not prove wholly devoid of interest even to those who are thoroughly familiar with the subject. Furthermore, the atomo-mechanical theory, which is supposed to be the only and all-sufficient basis of the science of physics, has become complicated with, or, rather, has led to, certain remarkable speculations as to the nature and properties of space; and this necessitated another excursion into the field of mathematics, for the purpose of examining the validity of the doctrines of what is generally known as transcendental geometry with its hypotheses of non-homaloidal space and of space of more than three dimensions.

What is here presented is not, of course, a new theory of the universe, or a novel system of philosophy. I have undertaken, not to solve all or any of the problems of cognition, but simply to show that some of them are in need of being stated anew so as to be rationalized, if not deepened. It is an old truth, which, however, is too often lost sight of, that many of the questions of science and philosophy remain unanswered, not by reason of the insufficiency of our knowledge, but because the questions themselves are founded on erroneous assumptions and require answers in irrational or impossible terms. The utter anarchy which notoriously prevails in the discussion of ultimate scientific questions, so called, indicates that a determination of the proper attitude of scientific inquiry toward its objects is the most pressing intellectual need of our time, as it is an indispensable prerequisite of real intellectual progress at all times. And such a determination, however partial, is in itself a decided advance in the direction of our legitimate cognitive aspirations. "Rightly to propose a problem," says Whewell, "is no inconsiderable step to its solution." In the language of Kant: "*Es ist schon ein grosser und noethiger Beweis der Klugheit und Einsicht zu wissen, was man vernuenftiger Weise fragen solle.*" And in the pithy phrase of Bacon: "*Prudens interrogatio quasi dimidium scientiae.*"

My views respecting the actual state of physical science and the value of many of the current theoretical interpretations of scientific facts are, no doubt, at variance with the tenets of many distinguished scientific men. That I have, nevertheless, given fearless expression to them will not, I hope, be construed as a want of appreciation of the merits of those to whose labors modern culture owes its life, and the pursuit of knowledge in the interest of that culture its practical success. And, if it should be regarded as evidence of presumption, I desire to say that there are suggestions, in many of the utterances of the men of science here referred to, of a growing sense of the questionability of some of the elements of their scientific faith. I have taken frequent occasion, in the progress of my discussion, to point to these suggestions, to the end of showing that my thoughts are, after all, but the inevitable outcome of the tendencies of modern science, and are, therefore, rather "*partus temporis quam ingenii.*"

I deem it important to have it understood, at the outset, that this treatise is in no sense a further exposition of the doctrines of a book ("The Philosophy of Nature," Boston, Crosby & Nichols, 1848) which I published more than a third of a century ago. That book was written while I was under the spell of Hegel's ontological reveries—at a time when I was barely of age and still seriously affected with the metaphysical malady which seems to be one of the unavoidable disorders of intellectual infancy. The labor expended in writing it was not, perhaps, wholly wasted, and there are things in it of which I am not ashamed, even at this day; but I sincerely regret its publication, which is in some degree atoned for, I hope, by the contents of the present volume.

It ought to be added that parts of the seventh and eleventh chapters of this book, and a few sentences in the other chapters, were published in "The Popular Science Monthly" in October, November, and December, 1873, and January, 1874.

J. B. Stallo.

CONTENTS.

CHAPTER IX.

CHAPTER X.

CHAPTER XI.

CHAPTER XII.

CHAPTER XIII.

CHAPTER XIV.

CHAPTER XV.

CHAPTER XVI.

THE CONCEPTS AND THEORIES

OF

MODERN PHYSICS.

CHAPTER I.

INTRODUCTORY.

Modern physical science aims at a mechanical interpretation of all the phenomena of the universe. It seeks to explain these phenomena by reducing them to the elements of mass and motion and exhibiting their diversities and changes as mere differences and variations in the distribution and aggregation of ultimate and invariable bodies or particles in space. Naturally the supremacy of mechanics became conspicuous first in the domains of those sciences which deal with the visible motions of palpable masses—in astronomy and molar physics; but its recognition is now all but universal in all the physical sciences, including, not only molecular physics and chemistry, but also such departments of scientific inquiry as are conversant about the phenomena of organic life.

It is said that the theoretical no less than the practical progress of the natural sciences, during the last three centuries, is an achievement of mechanics which, besides devising the instruments of successful scientific research, has also supplied its principles and methods. It is, indeed, incontestable that the attempt at a consistent application of mechanical principles marks a new epoch in the history of science. The founders of modern physics proceeded upon the tacit if not upon the declared assumption that all true explanations of natural phenomena are mechanical explanations. That this did not at once find articulate expression is due, partly to the fact that principles are wont to assert themselves, in thought as in action, before they are distinctly apprehended, and partly to the circumstance that science, for a long time, was constrained to flourish under the shadows of metaphysics and theology. But it was not long after the days of Stevinus, Fermat and Galilei before the doctrine that all physical action is mechanical was stated in terms. Even during the life of Galilei—a year before his death—Descartes announced that "all variations of matter, or all diversity of its forms, depends on motion." * And nine years before the appearance of Newton's *Principia* Thomas Hobbes declared that "change [i. e., physical change] is of necessity nothing else than motion of the parts of the body changed," † at the same time adding that "there can be no cause of motion in a body but in another body contiguous and moved." ‡ Leibnitz was even more emphatic, asserting that the doctrine in question is not merely an experiential induction, but a self-evi-

* "Omnis materiae variatio sive omnium ejus formarum diversitas pendet a motu." Cartes. Princ. Phil. ii, 23.

† "Necesse est ut mutatio aliud non sit praeter partium corporis mutati motum." Hobbes, Philos. prima, pars secunda, ix, 9.

‡ "Causa motûs nulla esse potest in corpore nisi contiguo et moto." *Ib.*

dent truth. "Everything in nature," he said, "is effected mechanically—a principle which can be made certain by reason alone, and never by experiments, however numerous they may be." * He, too, insisted that all motion is caused by impact. "A body is never moved naturally, except by another body which presses in touching it." † Similarly Huygens, the great contemporary of Leibnitz and Newton, said that "in true philosophy the causes of all natural effects are, and in his judgment must be, conceived mechanically, unless we are to renounce all hope of understanding anything in physics." ‡ And in the first comprehensive treatise on physics ever published, that of Musschenbroek, it is put forth as an axiom that "no change is induced in bodies whose cause is not motion." #

The most definite statement, however, of the proposition that the true aim and object of all physical science is a reduction of the phenomena of nature to a coherent mechanical system is found in the scientific

* "Tout se fait mécaniquement dans la nature, principe qu'on peut rendre certain par la seule raison et jamais par les expériences, quelque nombre qu'on en fasse." Leibnitz, Nouveaux Essais, Opp. ed. Erdmann, p. 383.

† "Un corps n'est jamais mû naturellement que par un autre corps qui le presse en le touchant." 5me lettre à Clarke, Erdmann, p. 767. Hence Wolff, the dogmatic expounder of the Leibnitian philosophy: "Corpus non agit in alterum nisi dum in ipsum impingit." Wolff, Cosmologia gen., 129.

‡ " . . . in verâ philosophiâ, in quâ omnium effectuum causae concipiuntur per rationes mechanicas: id quod meo judicio fieri debet nisi velimus omnem spem abjicere aliquid in physicis intelligendi." Hugenii Opp. reliqua, Amst., 1728, vol. i (Tract. de lumine), p. 2.

"Nulla autem corporibus inducitur mutatio, cujus causa non fuerit motus, sive excitatus, sive minutus, aut suffocatus; omne enim incrementum vel decrementum, generatio, corruptio, vel qualiscunque alteratio, quae in corporibus contingit, a motu pendet." P. v. Musschenbroek, Introd. ad. philos. naturalem, vol. i, cap. 1, § 18 (ed. Patav., 1768).

writings published during the second half of the present century, since the discoveries made in organic chemistry by the aid of the atomic theory, the revelations of the spectroscope, the establishment of the doctrine of the conservation of energy, and the promulgation of the mechanical theory of heat with its complement, the kinetic theory of gases. Thus Kirchhoff, one of the founders of the theory of spectral analysis, said in 1865: "The highest object at which the natural sciences are constrained to aim, but which they will never reach, is the determination of the forces which are present in nature, and of the state of matter at any given moment—in one word, the reduction of all the phenomena of nature to mechanics." * To the same effect Helmholtz, in his inaugural address delivered before the meeting of the association of physicians and naturalists at Innspruck, in 1869: "The object of the natural sciences is to find the motions upon which all other changes are based, and their corresponding motive forces—to resolve themselves, therefore, into mechanics." † No less pointed are the words of Clerk Maxwell: "When a physical phenomenon," he writes, "can be completely described as a change in the configuration and motion of a material system, the dynamical explanation of that phenomenon is said to be com-

* "Das hoechste Ziel, welches die Naturwissenschaften zu erstreben haben, aber niemals erreichen werden, ist die Ermittelung der Kraefte, welche in der Natur vorhanden sind und des Zustandes in dem die Materie in einem Augenblick sich befindet, mit einem Worte, die Zurueckfuehrung aller Naturerscheinungen auf die Mechanik." Kirchhoff, Ueber das Ziel der Naturwissenschaften. Prorectoratsrede, Heidelberg, 1865. S. 9, 24.

† "Das Endziel der Naturwissenschaften ist, die allen andern Veraenderungen zu Grunde liegenden Bewegungen und deren Triebkraefte zu finden, also sich in Mechanik aufzuloesen." Helmholtz, Populaerwissenschaftliche Vortraege, i, 93.

plete. We can not conceive any further explanation to be either necessary, desirable, or possible, for as soon as we know what is meant by the words *configuration*, *mass* and *force*, we see that the ideas which they represent are so elementary that they can not be explained by means of anything else." *

Citations like these, from the writings of eminent physicists, might be multiplied almost indefinitely. And, if we turn from the physicists to the physiologists, we meet with declarations equally explicit. "Every analysis," said Ludwig in 1852, "of the animal organism has thus far brought to light a limited number of chemical atoms, the presence of the light- (heat-) bearing æther and of the electric fluids. These data lead to the inference that all the phenomena of animal life are consequences of the simple attractions and repulsions resulting from the concurrence of these elementary substances." † In a similar strain Wundt, writing twenty-five years later: "The view that has now become dominant [in physiology], and is ordinarily designated as the mechanical or physical view, has its origin in the causal conception long prevalent in the kindred departments of natural science, which regards nature as a single chain of causes and effects wherein the ultimate laws of causal action are the laws of mechanics.

* Clerk Maxwell, "On the Dynamical Evidence of the Molecular Constitution of Bodies." "Nature," March 4 and 11, 1875.

† "So oft nun eine Zergliederung der leistungserzengenden Einrichtungen des thierischen Koerpers geschah, so oft stiess man schliesslich auf eine begrenzte Zahl chemisher Atome, die Gegenwart des Licht- (Waerme-) Aethers und diejenige der electrischen Fluessigkeiten. Dieser Erfahrung entsprechend zieht man den Schluss, dass alle vom thierishen Koerper ausgehenden Ersheinungen eine Folge der einfachen Anziehungen und Abstossungen sein moechten, welche an jenen elementaren Wesen bei einem Zusammentreffen derselben beobachtet werden." Ludwig, Lehrbuch der Physiologie des Menschen, Band i, Einleitung, p. 2.

Physiology thus appears as a branch of applied physics, its problem being a reduction of vital phenomena to general physical laws, and thus ultimately to the fundamental laws of mechanics." * Still more broadly, Haeckel: "The general theory of evolution . . . assumes that in nature there is a great, unital, continuous and everlasting process of development, and that all natural phenomena without exception, from the motion of the celestial bodies and the fall of the rolling stone up to the growth of the plant and the consciousness of man, are subject to the same great law of causation—that they are ultimately to be reduced to atomic mechanics." † This theory, Haeckel declares, "is the only scientific theory which affords a rational explanation of the universe, and satisfies the craving of the intellect for causal connections, inasmuch as it links all the phenomena of nature as parts of a great unital process of development and as a series of mechanical causes and

* "Die jetzt zur Herrschaft gelangte Auffassung dagegen, die man als die physikalishe oder mechanistische zu bezeichnen pflegt, ist aus der in den verwandten Zweigen der Naturwissenschaft schon laenger zur Geltung gekommenen causalen Naturansicht entsprungen, welche die Natur als einen einzigen Zusammenhang von Ursachen und Wirkungen ansieht, wobei als letzte Gesetze, nach denen die natuerlichen Ursachen wirken, sich stets die Grundgesetze der Mechanik ergeben. Die Physiologie erscheint daher als ein Zweig der angewandten Naturlehre. Ihre Aufgabe erkennt sie darin, die Lebenserscheinungen auf die allgemeinen Naturgesetze, also schliesslich auf die Grundgesetze des Mechanik, zurueckzufuehren." Wundt, Lehrbuch der Physiologie des Menschen, 4te Auflage, p. 2.

† "Die allgemeine Entwickelungslehre . . . nimmt an, dass in der ganzen Natur ein grosser, einheitlicher, ununterbrochener und ewiger Entwickelungsvorgang stattfindet, und dass alle Naturerscheinungen ohne Ausnahme, von der Bewegung der Himmelskoerper und dem Fall des rollenden Steins bis zum Wachsen der Pflanze und zum Bewusstsein des Menschen, nach einem und demselben grossen Causal-Gesetze erfolgen, dass alle schliesslich auf Mechanik der Atome zurueckzufuehren sind." Haeckel, Freie Wissenschaft und freie Lehre, pp. 9, 10.

effects." * In the same sense Huxley speaks of "that purely mechanical view toward which modern physiology is striving." †

A very lucid and thorough exposition of the aims of modern physical science is contained in the following passage taken from a recent lecture of Emil Du Bois-Reymond—equally distinguished as a physicist and physiologist: "Natural science—more accurately expressed, scientific cognition of nature, or cognition of the material world by the aid and in the sense of theoretical physical science—is a reduction of the changes in the material world to motions of atoms caused by central forces independent of time or a resolution of the phenomena of nature into atomic mechanics. It is a fact of psychological experience that, whenever such a reduction is successfully effected, our craving for causality is, for the time being, wholly satisfied. The propositions of mechanics are reducible to mathematical form, and carry within them the same apodictic certainty which belongs to the propositions of mathematics. When the changes in the material world have been reduced to a constant sum of potential and kinetic energy inherent in a constant mass of matter, there is nothing left in these changes for explanation.

"The assertion of Kant, in the preface to the 'Metaphysical Rudiments of Natural Science,' that 'in every department of physical science there is only so much science, properly so called, as there is mathematics,' is

* "Der Monismus, die universale Entwickelungstheorie, oder die monistische Progenesistheorie ist die einzige wissenschaftliche Theorie, welche das Weltganze vernunftgemass erklaert, und das Causalitaetsbeduerfniss unserer menschlichen Vernunft befriedigt, indem sie alle Natur-Erscheinungen als Theile eines einheitlichen grossen Entwickelungs-Processes in mechanischen Causal-Zusammenhang bringt." *Ibid.*, p. 11.

† Lay Sermons, Addresses and Reviews (Appletons' ed.), p. 331.

to be sharpened by substituting 'mechanics of atoms for 'mathematics.' This was evidently his own meaning when he denied the name 'science' to chemistry. It is not a little remarkable that in our time chemistry, since it has been constrained, by the discovery of substitution, to abandon the old electro-chemical dualism, has seemingly taken a retrograde step in its advance toward science in this sense. *The resolution of all changes in the material world into motions of atoms caused by their constant central forces would be the completion of natural science.*" *

* "Naturerkennen—genauer gesagt, naturwissenschaftliches Erkennen oder Erkennen der Koerperwelt mit Huelfe und im Sinne der theoretischen Naturwissenschaft—ist Zurueckfuehren der Veraenderungen in der Koerperwelt auf Bewegungen von Atomen die durch deren von der Zeit unabhaengige Centralkraefte bewirkt werden, oder Aufloesung der Naturvorgaenge in Mechanik der Atome. Es ist psychologische Erfahrungsthatsache, dass wo solche Aufloesung gelingt, unser Causalitaetsbeduerfniss vorlaeufig sich befriedigt fuehlt. Die Saetze der Mechanik sind mathematisch darstellbar, und tragen in sich dieselbe apodiktische Gewissheit, wie die Saetze der Mathematik. Indem die Veraenderungen in der Koerperwelt auf eine constante Summe potentieller und kinetischer Energie, welche einer constanten Menge von Materie anhaftet, zurueckgefuehrt werden, bleibt in diesen Veraenderungen selber nichts zu erklaeren uebrig.

"Kant's Behauptung in der Vorrede zu den 'Metaphysischen Anfangsgruenden der Naturwissenschaft,' 'dass in jeder besonderen Naturlehre nur so viel eigentliche Wissenschaft angetroffen werden koenne, als darin Mathematik anzutreffen sei,' ist also vielmehr noch dahin zu verschaerfen, dass fuer Mathematik Mechanik der Atome gesetzt wird. Sichtlich diess meinte er selber als er der Chemie den Namen einer Wissenschaft absprach, und sie unter die Experimentallehren verwies. Es ist nicht wenig merkwuerdig dass in unserer Zeit die Chemie indem sie durch die Entdeckung der Substitution gezwungen wurde den electrochemischen Dualismus aufzugeben, sich von dem Ziel, eine Wissenschaft in diesem Sinne zu werden, scheinbar wieder weiter entfernt hat. Denken wir uns alle Veraenderungen in der Koerperwelt in Bewegungen von Atomen aufgeloest, die durch deren constante Centralkraefte bewirkt werden, so waere das Weltall naturwissenschaftlich erkannt." Emil Du Bois-Reymond, "Ueber die Grenzen des Naturerkennens," p. 2 *seq.*

With few exceptions, scientific men of the present day hold the proposition, that all physical action is mechanical, to be axiomatic, if not in the sense of being self-evident, at least in the sense of being an induction from all past scientific experience. And they deem the validity of the mechanical explanation of the phenomena of nature to be, not only unquestionable, but absolute, exclusive, and final. They believe that this validity is not conditioned, either by the present state of human intelligence, or by the nature and extent of the phenomena which present themselves as objects of scientific investigation. Thoughtful men like Du Bois-Reymond have at times suggested that it is not unlimited ; but the only limits assigned to it are those of the general capacity of the human intellect. Although they concede that there is a class of phenomena—those of organic life — which, under their characteristic aspect, are wholly irreducible by the mere aid of mechanical principles, it is, nevertheless, insisted that these principles constitute the only intellectual solvent that can be applied to them, and that the residue which resists the solution is to be relegated for ever to that endless array of facts which are proof against all the reagents of scientific cognition. It is claimed that, if it is impossible theoretically to construct a living organism out of molecules or atoms, and mechanical forces under the guidance of the principle of the conservation of energy, the laws of electric or magnetic coercion, the first and second laws of thermo-dynamics, etc., the attempt to frame a theory of life in harmony with the laws controlling ordinary material action must be utterly abandoned. Such a claim ought not, in my judgment, to be admitted without a careful examination of the grounds upon which it is made. It is my purpose, therefore, in the following pages to inquire whether or not the validity of the mechanical theory of the universe in its present form, and with its ordinary assumptions, is indeed absolute within the bounds of human intelligence, and to this end, if possible, to ascertain the nature of this theory as well as its logical and psychological origin. Obviously the first question presenting itself in the course of an examination into its validity is whether it is consistent with itself and with the facts for the explanation of which it is propounded. Our initial problem, then, will be that of finding an answer to this question.

CHAPTER II.

FIRST PRINCIPLES OF THE MECHANICAL THEORY OF THE UNIVERSE.

THE mechanical theory of the universe undertakes to account for all physical phenomena by describing them as variances in the structure or configuration of material systems. It strives to apprehend all phenomenal diversities in the material world as varieties in the grouping of primordial units of mass, to recognize all phenomenal changes as movements of unchangeable elements, and thus to exhibit all apparent qualitative heterogeneity as mere quantitative difference. In the light of this theory the ultimates of scientific analysis are *mass* * and *motion*, which are assumed to be essentially disparate. Mass, it is said, exists independently of motion and is indifferent to it. It is the same whether it be in motion or at rest. Motion may be transferred from one mass to another without destroying the identity of either.

The prime postulate of all science is that there is some constant amid all phenomenal variations. Science is possible only on the hypothesis that all change is in its nature transformation. Without this hypothesis it

* It is hardly necessary to say that I purposely designate *mass*, and not (as is usual) *matter*, as the correlate of motion. When a body is divested, in thought, of all those qualities which, according to the teachings of modern science, are in their nature phases of motion, the residue is not matter, but mass.

could discharge neither of its two great functions—those of determining, from the present state of things, the past on the one hand and the future on the other, by exhibiting the one as its necessary antecedent, and the other as its equally necessary consequent. It is evident that the computations of science would be utterly frustrated by the sudden disappearance of one or more of its elements, or the unbidden intrusion of new elements. If, therefore, scientific analysis yields mass and motion as its absolutely irreducible elementary terms—if these terms underlie all possible transformations—it follows that both are quantitatively invariable. Accordingly the mechanical theory of the universe postulates the conservation of both mass and motion. Mass may be transformed by an aggregation or segregation of parts; but amid all these transformations it persistently remains the same. Similarly motion may be distributed among a greater or less number of units of mass; it may be transferred from one unit of mass to any number of units, its velocity being reduced in proportion to the number of units to which the transference takes place; nevertheless the sum of the motions of the several units is always equal to the motion of the single unit. It may be changed in direction and form; rectilinear motion may become curvilinear, translatory motion may be broken up into vibratory motion, molar motion may be converted into molecular agitation; yet, during all these changes, it is never increased, diminished, or lost. The conservation of mass (or, as it is generally but inaccurately termed, the conservation or indestructibility of matter) has long been a standing axiom of physical science. The conservation of motion (i. e., the conservation of energy, which, as will hereafter appear, is, according to the mechanical theory, the same thing), though but recently formulated as a distinct scientific principle, is now universally regarded as of equal evidence and axiomatic dignity with its older counterpart. Indeed, while chemistry is said to be founded on the conservation of matter,* the recent progress of theoretical physics has consisted mainly in the solution of the problem of reconstituting it on the basis of the conservation of energy. The science of physics, in addition to the general laws of dynamics and their application to the interaction of solid, liquid, and gaseous bodies, embraces the theory of those agents which were formerly designated as imponderables—light, heat, electricity, magnetism., etc.; and all these are now treated as forms of motion, as different manifestations of the same fundamental energy, and as controlled by laws which are simple corollaries from the law of its conservation. The only apparent exception is the second law of thermodynamics, a reduction of which, however, to the principle of least action, or rather Hamilton's extension of it, the principle of varying action, has been attempted by Boltzmann and Clausius, while others (among them Rankine, Szily, and Eddy) have sought to derive it directly from the principle of the conservation of energy.

It is thus seen that the theory according to which the cause of all phenomenal change and variety in nature is motion, and all apparent qualitative diversity is in reality quantitative difference, involves three propositions, which may be stated as follows:

I. *The primary elements of all natural phenomena*

* It is gradually coming to be understood that the conservation of energy is as important a principle in chemistry as that of the conservation of mass; but as yet chemical notation takes account of masses only and makes no exhibition of the quantities of energy gained or lost in any given chemical transformation.

—the ultimates of scientific analysis—are mass and motion.

II. *Mass and motion are disparate. Mass is indifferent to motion, which may be imparted to it, and of which it may be divested, by a transference of motion from one mass to another. Mass remains the same, whether at rest or in motion.*

III. *Both mass and motion are constant.*

Among the corollaries from the first and second of these propositions there are two which are as obvious as they are important: the inertia and the homogeneity of mass. Mass and motion being radically disparate, it is evident that mass can not be motion or the cause of motion—it is inert. And mass in itself can not be heterogeneous, for heterogeneity is difference, and all difference is caused by motion.

The propositions above set forth lie at the base of the whole mechanical theory. They command universal assent among physicists of the present day, and are to be regarded as the fundamental axioms of modern physical science. In addition to these propositions, however, there is the assumption, generally prevalent among physicists and chemists, of the molecular or atomic constitution of bodies, according to which mass is not continuous, but discrete, being an aggregate of unchangeable, and, in that sense at least, simple units. This assumption leads to four other propositions, which, in conjunction with the principle of the conservation of both mass and motion, may be said to constitute the foundations of the atomo-mechanical theory. They are these:

1. *The elementary units of mass, being simple, are in all respects equal.* This is manifestly nothing more than an assertion of the homogeneity of mass in con-

formity with the hypothesis of its molecular or atomic constitution.

2. *The elementary units of mass are absolutely hard and inelastic*—a necessary consequence of their simplicity, which precludes all motion of parts, and, therefore, all change of figure.

3. *The elementary units of mass are absolutely inert and therefore purely passive;* hence there can be no mutual action between them, other than mutual displacement caused by impulses from without.

4. *All potential energy, so called, is in reality kinetic.* Mass and motion being fundamentally disparate and inconvertible, and mass being absolutely inert, whatever be its position, motion can not originate in, or be caused by, anything but motion. Energy due to mere position, therefore, is impossible.

It is necessary to take up these propositions severally in their order, and to ascertain whether, and to what extent, they are consistent with, and serve as the explanation of, the facts of scientific experience.

CHAPTER III.

THE PROPOSITION THAT THE ELEMENTARY UNITS OF MASS ARE EQUAL.

If all the diversities in nature are caused by motion, it follows that mass, the substratum of this motion, is fundamentally homogeneous. This is so evident that, in the first distinct announcements of the mechanical theory, the two propositions—the principle and its corollary—appeared side by side. Thus the statement of Descartes cited in the first chapter * is accompanied by the declaration that "the matter which exists in the world is everywhere one and the same." † It is true that Descartes did not assert the absolute equality of single material elements, because he recognized but two primary properties of matter, extension and mobility, and therefore denied its atomic constitution. But, when in time the hypothesis of the atomic or molecular structure of matter became one of the cardinal doctrines of modern physical science, the postulate of the fundamental homogeneity of mass necessarily assumed the form of an assertion of the absolute equality of its primordial units. For reasons to be discussed presently, physicists, and especially chemists, of our day evince a disposition to ignore this essential feature of the mechanical

* Supra, p. 16.

† "*Materia itaque in toto universo una et eadem existit.*" Cart. Princ. Phil., ii, 23.

theory; but, among those who understand that all scientific theories must at last be brought to the test of consistency, it has rarely failed to meet with direct or implied recognition. "Chemistry," says Professor Wundt, "still refers the divergent qualities of matter to an original qualitative difference between the atoms. But the whole tendency of physical atomism is to derive all the qualitative properties of matter from the forms of atomic motion. *Thus the atoms themselves remain as elements utterly devoid of quality.*" * Of the same import are the words of Herbert Spencer: "The properties of the different elements result from differences of arrangement, arising by the compounding and recompounding of *ultimate homogeneous units.*" † Even in the writings of distinguished chemists there is no lack of utterances bearing testimony to the pressure of the logical necessity which constrains the modern physicist to insist upon the fundamental equality of the material elements. "It is conceivable," says Thomas Graham, "that the various kinds of matter now recognized as different elementary substances may possess one and the same ultimate or atomic molecule existing in different conditions of movement. The essential unity of matter is an hypothesis in harmony with the equal action of gravity upon all bodies. We know the anxiety with which this point was investigated by Newton and the care he took to ascertain that every kind of substance,

* Die abweichenden Eigenschaften der Materie verlegt die Chemie noch jetzt in eine urspruengliche qualitative Verschiedenheit der Atome. Nun geht offenbar die ganze Entwickelung der physikalischen Atomistik darauf aus, alle qualitativen Eigenschaften der Materie aus den Bewegungsformen der Atome abzuleiten. Die Atome selbst bleiben so nothwendig als volkommen qualitaetslose Elemente zurueck. "Die Theorie der Materie," Deutsche Rundschau, December, 1875, p. 381.

† Contemporary Review, June, 1872.

'metals, stones, woods, grain, salts, animal substances,' etc., are similarly accelerated in falling, and are therefore equally heavy.

"In the condition of gas, matter is deprived of numerous and varying properties with which it appears invested when in the form of a liquid or solid. The gas exhibits only a few grand and simple features. These again may all be dependent upon atomic or molecular mobility. Let us imagine one kind of substance only to exist—ponderable matter; and further, that matter is divisible into ultimate atoms, *uniform in size and weight.* We shall then have one substance and a common atom. With the atom at rest the uniformity of matter would be perfect. But the atom possesses always more or less motion, due, it must be assumed, to a primordial impulse. This motion gives rise to volume. The more rapid the movement the greater the space occupied by the atom, somewhat as the orbit of a planet widens with the degree of projectile velocity. Matter is thus made to differ only in being lighter or denser matter. The specific motion of an atom being inalienable, light matter is no longer convertible into heavy matter. In short, matter of different density forms different substances—different inconvertible elements as they have been considered.

"But, further, these more or less mobile, or light and heavy forms of matter, have a singular relation connected with equality of volume. Equal volumes of two of them can coalesce together, unite their movement, and form a new atomic group, retaining the whole, the half, or some simple proportion of the original movement and consequent volume. This is chemical combination. It is directly an affair of volume, and only indirectly connected with weight.

Combining weights are different, because the densities, atomic and molecular, are different." *

Views analogous to those of Graham are held by C. R. A. Wright, who suggests "that there is but one kind of primordial matter, all so-called elements and compounds being, as it were, allotropic modifications of this matter, differing from one another in the amount of energy latent per unit of mass."† And although Prout's conjecture, that the several chemical elements are really compounds or allotropic forms of hydrogen, has been definitively abandoned (even by Dumas and others who at divers times sought to revive it), it having been shown that the hypothesis, according to which the atomic weights of all the elements are exact multiples of that of hydrogen, is untenable, yet attention has lately been drawn to the fact that there seem to be spectroscopic indications of the predominance of a few gaseous elements, such as hydrogen and nitrogen, in certain nebulæ which appear to represent the earlier stages of planetary or stellar development, and of a gradual increase of metallic and other substances in more advanced forms—in other words, of a progressive differentiation of matter, a gradual advance from homogeneity to heterogeneity, on the successive stages of planetary or stellar evolution.‡

Now, while the absolute equality of the primordial units of mass is thus an essential part of the very foundations of the mechanical theory, the whole modern science of chemistry is based upon a principle directly

* "Speculative Ideas respecting the Constitution of Matter," Phil. Mag., 4th ser., vol. xxvii, p. 81 *seq.*

† Chemical News, October 31, 1873.

‡ Cf. F. W. Clarke, "Evolution and the Spectroscope," Popular Science Monthly, January, 1873, p. 320 *seq.* Lockyer's recent investigations have brought these views into great prominence.

subversive of it—a principle of which it has recently been said that "it holds the same place in chemistry that the law of gravitation does in astronomy." * This principle is known as the law of Avogadro or Ampère. It imports that equal volumes of all substances, when in the gaseous state and under like conditions of pressure and temperature, contain the same number of molecules—whence it follows that the weights of the molecules are proportional to the specific gravities of the gases; that, therefore, these being different, the weights of the molecules are different also; and, inasmuch as the molecules of certain elementary substances are monatomic (i. e., consist of but one atom each), while the molecules of various other substances contain the same number of atoms, that the ultimate atoms of such substances are of different weights.

The law of Avogadro, though, like all physical theories, an hypothesis, is believed to be the only hypothesis which is competent to account for the well-known variation of the volume of a gas inversely as the pressure (law of Boyle or Mariotte) and directly as the absolute temperature (law of Charles) as well as for the combination of gases in simple volumetric proportions (law of Gay-Lussac); and it has served as the basis of innumerable deductions respecting the formation and transformation of chemical compounds which have thus far met with unfailing experimental verification.

That this cardinal principle of modern theoretical chemistry is in utter and irreconcilable conflict with the first proposition of the atomo-mechanical theory is apparent at a glance. No reconciliation, certainly, is possible on the hypothesis suggested by Graham. For

* J. P. Cooke, The New Chemistry, p. 13.

that accounts for differences of density by attributing to equal primordial atoms unequal volumes resulting from their occupancy of unequal spaces by virtue of differences in the velocities of movement with which the several kinds of atoms are supposed to be inalienably endowed. It accounts for inequalities in the volumes of equal masses, not for inequalities of mass in equal volumes, and can not serve as an explanation of the latter, unless it is supplemented by the further assumption—to which, indeed, it lends little, if any, aid—that some, if not all, of the molecules are compounds or aggregates of different degrees of complexity. Two masses or molecules of equal volumes can be of different densities or weights only if the number of units contained in one is different from the number of units in the other. But Avogadro's law constrains the chemist to assume that the molecules of various elementary substances, notwithstanding the diversity of their weights, consist of the same number of atoms. Thus hydrogen and chlorine, whose molecular weights are two and seventy-one respectively, are both held to be diatomic, i. e., their molecules are held to consist of two atoms each. In the case of monads, or univalent elements, such as those just mentioned, the reasoning upon which this assumption rests is very simple. One volume of hydrogen combines with one volume of chlorine, forming two volumes of hydrochloric acid. Each volume of this compound, according to Avogadro's law, contains as many molecules as either volume of the constituent simple elements before combination; the two volumes of the compound, therefore, contain twice as many molecules as either volume of the constituents. But, in each molecule of the compound, both hydrogen and chlorine are present, whence it follows that each molecule of hydrogen, as well as each molecule of chlorine, must have contributed at least one atom to each molecule of hydrochloric acid, and thus must have consisted of at least two atoms.

The argument in the case of dyads (such as oxygen, sulphur, selenium, etc.), and other elements of still higher quantivalence, though somewhat less simple, is equally cogent upon the basis of Avogadro's law.

It may be said that the law in question determines only the minimum number of atoms in each molecule, leaving the maximum indeterminate, so that, after all, the molecule of greater weight may be of correspondingly greater complexity. But here we encounter an obstacle presented by a branch of the atomic theory in physics—the science of thermo-dynamics. Modern science regards heat as a form of energy—as consisting in an agitation of the molecules or atoms whereof bodies are composed; and, in the case of gaseous bodies at least, it discriminates between that part of this energy which is exhibited in the form of temperature, attributing it to translatory motions of the molecules, or rather of their centers of mass, and another part—the internal energy, so called—which is supposed to be dependent upon oscillatory or rotatory motions of their component atoms. It has been shown, experimentally, that the ratio of the specific heat of a gas at constant pressure to that at constant volume* falls short of the value assigned to it by the theory upon the supposition that all the heat imparted to a gaseous body is expended in producing a translatory motion of the molecules, the

* The "specific heat" (i. e. the heat required to raise the temperature of a unit of mass of any substance one degree) of a gas at constant pressure under which it expands, is necessarily greater than that at constant volume, because in the former case part of the heat is expended in the mechanical work of expansion.

effect being expansion, or increased pressure, or both; and this difference is accounted for by the assumption that part of the heat is converted into intramolecular agitation, i. e., into motions of the particles within the molecule which do not affect its position or action as a whole. Now, it is readily seen and has been shown by Clausius, Boltzmann, Maxwell, and others, that the energy thus converted into intramolecular or interatomic agitation must increase as the complexity of the molecular constitution increases; it would become enormous, therefore, if a molecule consisted of a number of atoms so great as to be sufficient to account for the differences between the molecular weights of the elements. The molecular weight of chlorine, for example, is 35·5 times as great as that of hydrogen; and if these weights are in proportion to the number of atoms contained in each molecule, it becomes necessary to assume—even granting that hydrogen is strictly diatomic—that each chlorine molecule is composed of no less than seventy-one atoms. But, if this assumption were valid, nearly all the heat imparted to chlorine would be absorbed, i. e., converted into internal energy, and its calculated specific heat would far exceed the amount ascertained by actual experiment.

There are thus difficulties not of a speculative, but of a purely physical and chemical nature, which render the indefinite multiplication of atoms within the molecule, so as to account for the diversity of molecular weights, wholly inadmissible. Several elementary substances are known to conform to Avogadro's law only on the supposition that they are monatomic. Among them is mercury, whose molecular weight coincides with its atomic weight as established by all the chemical tests applicable to it, including that of Dulong and

Petit's law. And it has been demonstrated by Kundt and Warburg * that the ratio of the specific heat of mercurial vapor at constant pressure to that at constant volume, as ascertained by experiment, is precisely equal to its value calculated upon the basis of the absolute simplicity of the mercurial molecule and of the non-absorption of any part of the heat in intramolecular action.

In view of all this there seems to be no escape from the conclusion that the claim, according to which modern physical science is throughout a partial and progressive solution of the problem of reducing all physical phenomena to a system of atomic mechanics, is very imperfectly, if at all, countenanced by the actual constitution of theoretical chemistry—that this science, which is peculiarly conversant about atoms and their motions, is founded upon propositions destructive of the very basis upon which alone a consistent superstructure of atomic mechanics can be reared. And there appears to be little ground for the hope that these propositions may be speedily abandoned; for, in the opinion of the most distinguished chemists of the day, such an abandonment would throw the mass of chemical facts, laboriously ascertained by experiment and observation (induced, partly at least, by the propositions in question) into a state of hopeless prescientific confusion.

In reference to the speculations of those who seek to deduce the specific differences between the ultimate units of mass from differences between their supposed inalienable velocities of motion or amounts of latent energy, it is to be said, not only that they fail to afford a solution of the difficulties of theoretical chemistry in

* Pogg. Ann., vol. clvii, p. 353.

the presence of the inexorable demands of the mechanical theory, but also that the attribution of inalienable energy or motion to a given mass is repugnant to the fundamental postulate of the absolute indifference of mass to motion. Helmholtz and others have investigated the conditions of vortex motion in a perfectly homogeneous, incompressible and frictionless fluid, which (as Maxwell has shown) is of necessity continuous and can not be molecular or atomic. If these conditions could be realized, we should have constant but undistinguishable volumes of a permanently homogeneous fluid, so called, endowed with constant quantities of inalienable motion. But no energy or motion can inhere essentially in distinct and separate masses (molecules or atoms) if, as the mechanical theory assumes, mass and motion are disparate—if mass is indifferent to motion so as to remain the same whether in motion or at rest, and if motion is transferable from one mass to another. This is one of the points distinctly insisted upon by Sir Isaac Newton, the greatest among the founders of the mechanical theory. Newton distinguishes between two kinds of force—the force of inertia (*vis inertiae*), and impressed force (*vis impressa*). The former alone according to him is *vis insita*, i. e., inheres in matter; while of the latter he expressly says that "this force consists in action alone and does not abide in the body after action." *

* "*Consistit haec vis in actione solâ, neque post actionem permanet in corpore.*" Phil. Nat. Princ. Math., def. iv (éd. Le Seur et Jacquier, vol. i, p. 4).

CHAPTER IV.

THE PROPOSITION THAT THE ELEMENTARY UNITS OF MASS ARE ABSOLUTELY HARD AND INELASTIC.

FROM the essential disparity of mass and motion and the simplicity of the elementary units of mass it follows that these units are perfectly hard and inelastic. Elasticity involves motion of parts and can not, therefore, be an attribute of truly simple atoms. "The concept 'elastic atom,'" justly observes Professor Wittwer, "is a contradiction in terms, because elasticity presupposes parts the distances between which can be increased and diminished." *

The early founders of the mechanical theory regarded the absolute hardness of the component particles of matter as an essential feature of the original order of nature. "It seems probable to me," says Sir Isaac Newton, "that God in the beginning formed matter in solid, massy, hard, impenetrable, movable particles of such sizes and figures, and with such other properties and in such proportion to space as most conduced to the end for which he formed them; and that these primitive particles being solids are incomparably harder than any porous bodies compounded of them; even so

* "Der Begriff 'elastisches Atom' ist eine *contradictio in adjectis*, da die Elasticitaet immer wieder Theile voraussetzt, die sich einander naehern, die sich von einander entfernen koennen." Beitraege zur Molecularphysik, Schloemilch's Zeitschrift fuer Math. und Phys., vol. xv, p. 114.

very hard as never to wear or break in pieces; no ordinary power being able to divide what God himself made one in the first creation." *

Strangely enough, while the requirement, by the mechanical theory, of the absolute rigidity of the elementary units of mass is no less imperative than that of their absolute simplicity, it meets with an equally signal denial in modern physics. The most conspicuous among the hypotheses which have been devised since the general adoption of the modern theories of heat, light, electricity and magnetism, and the establishment of the doctrine of the conservation of energy, in order to afford consistent ground for the mechanical interpretation of physical phenomena, is that known as the kinetic theory of gases. In the light of this theory a gaseous body is a swarm of innumerable solid particles incessantly moving about with different velocities in rectilinear paths of all conceivable directions, the velocities and directions being changed by mutual encounters at intervals which are short in comparison with ordinary standards of duration, but indefinitely long as compared with the duration of the encounters. It is readily seen that these motions would soon come to an end if the particles were wholly inelastic, or imperfectly elastic. For in that case there would be loss of motion at every encounter. The assumed perpetuity of the motion of the particles, therefore, leads to the necessity of asserting their perfect elasticity. And this necessity results, not merely from the peculiar exigencies of the kinetic theory of gases, but also from the principle of the conservation of energy in its general application to the ultimate constituents of sensible masses, if these constituents are supposed to be in motion. In the case

* Opticks, fourth ed., p. 375.

of the collision of ordinary inelastic or partially elastic bodies there is a loss of motion which is accounted for by the conversion of the motion thus lost into an agitation of the minute parts composing the colliding bodies. But in atoms or molecules destitute of parts no such conversion is possible, and hence we are constrained to assume that the ultimate molecules of a gaseous body are absolutely elastic.

The necessity of attributing perfect elasticity to the elementary molecules or atoms in view of the kinetic theory of gases has been expressly recognized by all its founders. "Gases," says Kroenig,* "consist of atoms which behave like solid, *perfectly elastic* spheres moving with definite velocities in void space." This statement is adopted by Clausius † and emphasized by Maxwell, the first part of whose essay, "Illustration of the Dynamical Theory of Gases," is a treatise "on the motions and collisions of *perfectly elastic* spheres." ‡ And the highest scientific authorities are equally explicit in declaring that the hypothesis of the atomic or molecular constitution of matter is in conflict with the doctrine of the conservation of energy, unless the atoms or molecules are assumed to be perfectly elastic. "We are forbidden," says Sir William Thomson,§ "by the modern theory of the conservation of energy to assume inelasticity or anything short of perfect elasticity of the ultimate molecules, whether of ultra-mundane or mundane matter."

Naturally, eminent advocates of the kinetic hypothesis have taxed their ingenuity in the search of

* Pogg. Ann., vol. xcix, p. 316.
† *Ib.*, vol. c, p. 353.
‡ Phil. Mag., 4th ser., vol. xix, p. 19.
§ *Ib.*, vol. xlv, p. 321.

methods for the extrication of the mechanical theory from the dilemma in which it is thus involved. The most notable effort thus far made is that of Sir William Thomson, in the form of a conjecture suggested by the researches of Helmholtz,* respecting the properties of rotational motion in an absolutely homogeneous, incompressible, perfect fluid, to which reference has already been made in the preceding chapter. Thomson imagines the omnipresence of this fluid, and supposes that atoms are in fact vortex-rings formed by rotational movements within it. Such rings would be permanent, of invariable volume due to an invariable quantity of motion, though susceptible of a great variety of form; and some of their features, such as their modes of implication, would be indestructible; they would be capable of being knotted on themselves or linked with other vortex-rings, but could never be unknotted or untied; finally they would be incapable of interpenetration or coalescence, and their mutual approaches would result in rebounds similar to the resilience of perfectly elastic bodies.

While we willingly yield our homage to the sagacity displayed in this attempt to relieve the mechanical theory from one of its most fatal embarrassments, it is to be feared that its success is altogether illusory. For, it seems to be evident that motion in a perfectly homogeneous, incompressible and therefore continuous fluid is not sensible motion. All partition of such a fluid is purely ideal; in spite of the displacement of any portion of it by another portion, a given space would at any moment present the same quantity of substance absolutely indistinguishable from that present there a

* Cf. Crelle-Borchardt's Journal fuer reine und angewandte Mathematik, vol. lv, p. 25.

moment before. There would be no phenomenal difference or change. A fluid both destitute and incapable of difference is as impossible a vehicle of real motion as pure space; it is as useless for the purpose of accounting for the phenomena of material action as the quasi-material medium without inertia of which Roger Cotes said that it was not to be distinguished from a vacuum.*

Again, as Maxwell has observed,† the vortex-ring atoms moving in the hypothetical fluid would lack the essential attribute of matter: inertia. Such atoms would consist, not in the substance of the omnipresent fluid, but simply in the motions induced therein. Of these motions the persistence of both mass and energy would have to be predicated, and from them the concretions of mass, together with all the phenomena exhibited by sensible matter, would have to be derived. But that is impossible. From its very nature motion can not be the bearer of motion, nor can it, by itself, be the generator of momentum which is essentially the product of two antagonistic factors, and which would be utterly extinguished by the suppression of either. Upon the basis of the mechanical theory, the fundamental antithesis between mass and motion, inertia and energy, can not be destroyed without an obliteration of all the distinctions which constitute the elements of our conceptions respecting the nature of physical action.

Another attempt, somewhat analogous to that of Sir William Thomson, to dispense with the necessity of en-

* "Qui coelos materiâ fluidâ repletos esse volunt, hanc vero non inertem esse statuunt, hi verbis tollunt vacuum, re ponunt. Nam cum hujusmodi materia fluida ratione nullâ secerni possit ab inani spatio; disputatio tota fit de rerum nominibus, non de naturis. Praef. in Newtoni Phil. Nat. Princ. Math., ed. Le Seur et Jacquier, p. 25.

† Encycl. Brit., ninth ed., s. v. Atom.

dowing the elementary atoms with the intrinsic property of elasticity has been made by A. Secchi. This distinguished physicist and astronomer also derives the resilience of the ultimate particles from their rotatory motion; but his atoms, unlike those of Thomson, are real corpuscles separated by wide interstitial spaces, and not mere movements in a continuous and incompressible æthereal medium. Secchi clearly apprehends the inadmissibility of attributing elasticity to simple elementary atoms. "It is evident," he says,* "that, while it is possible to admit its existence in a compound molecule, the same thing can not be done in the case of elementary atoms. Indeed, elasticity in the received sense presupposes void spaces in the interior of the molecule whose form is changed by compression so as to return, afterward, to its original figure. Now, we regard the atoms as impenetrable, and not as groups of solid particles; hence they can not include void spaces which permit their dilatation and contraction.

"In truth, what we call a molecule of a simple (i. e., chemically undecomposable) gas is not an elementary atom, or at least is not necessarily one. Inasmuch as this gaseous molecule is an aggregate of veritable atoms, it may well be that it has internal pores, and, generally, a number of properties which do not belong to its constituent atoms; it is not absurd, therefore, to suppose it to be endowed with elasticity. Huygens has admitted this hypothesis for the æther. In his opinion the æthereal particles are composed of smaller ones; but on closer examination it is seen that this is a mere shifting of the difficulty, and not a solution of it. We hope to be able to show that it is nowise necessary to accept such an elasticity as a primitive force, and that

* L'unité des forces physiques, 2me éd., p. 47 *seq.*

the apparent repulsion of the atoms and their reciprocal collisions can be simply referred to an appropriate motion, it being sufficient for this purpose to suppose them to be in rotation. Let us prove this:

"Among the beautiful theorems discovered by Poinsot respecting the impact of bodies in rotation is found one relating to their reflection from a resisting obstacle. It teaches us that by virtue of its rotation alone a hard and inelastic body can rebound absolutely like a body perfectly elastic; more than that: one of these bodies, thrown against a fixed obstacle, is often sent back with a velocity superior to its initial velocity. The profound mathematician shows how this phenomenon, paradoxical as it seems, is due to the transformation of part of its rotatory motion into motion of translation; whence results an increase of the velocity of the center of gravity. According to the ordinary theories of impact, in which no account is taken of the motion of rotation, the preceding proposition is absurd, and nevertheless it is perfectly established. Thus, by the side of cases of ordinary *reflection* we find the phenomena of *progression;* we might also, using the expression of Poinsot, call them *negative reflections.*

"In negative reflection after impact, the center of gravity of the body returns with a velocity superior to that which it had at first. These questions form a wholly new and very interesting branch of mechanics; they are easily demonstrated by compounding the two movements of rotation and translation, considered with reference to the centers of gravity, of rotation and of percussion; and we readily understand that *generally* it may be said: an impact, whatever it may be, can never simultaneously annihilate in a body the two motions of rotation and of translation; for, when the impact is eccentric, it can destroy rotation and not translation, and, when the direction of the impact passes through the center of gravity, it can annihilate translation, but not rotation. Thus, the quantity of motion lost on the one side is gained on the other; the rotation may either be reversed or simply accelerated, according to the point of the body which is struck; whence the notion of *centers of conversion.* Examples of reflection succeeding the impact of bodies in rotation are found in the movements of disks and quoits, the impact of spinning-tops, etc. Billiard-players know perfectly how the rotation of the balls modifies the laws relating to the impact of elastic bodies as established in the elementary treatises." *

Unfortunately, the theory thus advanced finds little support in Poinsot's theorems. Secchi maintains that the impact of a rotating body, when it is eccentric, "can destroy rotation, but not translation," and, when its direction passes through the center of gravity, "it can annihilate translation, but not rotation," so that in either case "the quantity of motion lost on the one side is gained on the other." † But, from a careful examination of Poinsot's memoir, it appears that, after the colli-

* The theorems to which Secchi refers are contained in the last of a series of memoirs (Questions Dynamiques sur la Percussion des Corps) contributed by M. Poinsot to Liouville's Journal de Mathématiques pures et appliquées, 2me série, t. ii (1857), p. 281 *seq.*, and t. iv (1859), p. 421 *seq.* This remarkable memoir was published (and probably written) by the octogenarian geometer shortly before his death; the last installment, indeed, was published after his death in the same number of Liouville's Journal which contained the addresses pronounced at his funeral by MM. Bertrand and Mathieu.

† Secchi invariably speaks of loss or gain of "quantity of motion"; but his argument requires that this should be interpreted as meaning loss or gain of energy. Whether or not this is his own meaning, I do not undertake to say.

sion of rotating inelastic bodies, their rotation, or translation, or both are conserved, or the increase, diminution or loss of the one is compensated for by the diminution, increase or gain of the other, only in certain special cases. Poinsot shows * that, when a rotating inelastic body encounters a fixed obstacle, it depends on the distance between the spontaneous center of rotation and the center of gravity whether the body shall be reflected with a translatory velocity greater than, equal to, or less than its initial velocity, or shall lose its translatory velocity altogether. In the first place, there are always, between the center of gravity and the center of percussion, "two points such that, if the rotating body strikes the obstacle in the line of either, its center of gravity will be reflected with an increased velocity." † In the second place, "there are always, in every advancing rotating body, two points of perfect reflection, i. e., two points such that, if the body strikes an obstacle in the line of either, it will be reflected with a velocity perfectly equal to the velocity with which it is animated," ‡ so that "the center of gravity of the body is reflected in space as though the body were perfectly elastic." *But, when this occurs, the body loses, in the one case one third, and in the other two thirds of its velocity of rotation.*# Finally, in the third case, "if the obstacle is presented, either to the center of gravity or to the center of percussion, the velocity of translation is equally destroyed, the only difference between the two cases being that in the first case only the velocity of translation is destroyed without alteration of the velocity of rotation; while in the

* Liouville, Journal, etc., 2me série, t. ii, p. 288 *seq.*

† *L. c.*, p. 304. ‡ *L. c.*, p. 305.

\# *L. c.*, p. 307.

second case both the velocity of translation and the velocity of rotation are annihilated." *

The truth is, therefore, that in the only instances of perfect reflection specified by Poinsot there is a loss of either one third or two thirds of the rotatory motion not compensated for by any increase of translatory motion, and that there are cases of impact in which both the motion of translation and that of rotation simultaneously disappear.†

That Secchi should have deemed it possible to devolve the duty of conserving the energy of colliding atoms upon rotation as a substitute for the "occult quality" of perfect elasticity, seems almost incredible when we come to consider the use he makes of his own theory. This theory, according to him, serves as an explanation of a number of things, among which are the formation of molecular aggregates from simple atoms, and the phenomena of gravitation. The aggregation of atoms, so as to form compound molecules, he explains thus: ‡ "Suppose an extreme case, viz., the collision of

* *L. c.*, p. 308.

† Although I have long since become utterly indifferent to questions and claims of priority, it may not be improper to say here that the foregoing pages were written before I had seen the very able pamphlet "Das Raethsel der Schwerkraft" (Braunschweig, Vieweg und Sohn, 1879), of D. C. Isenkrahe, with whom I am happy to find myself in accord as to the validity of Secchi's attempt to deduce the property of perfect resilience from the rotation of inelastic bodies by the aid of Poinsot's exposition of the theory of rotation, although I can not, of course, accede to Isenkrahe's own theory of gravitation. There are other coincidences—all the more interesting because they are, no doubt, wholly accidental—between the criticisms contained in this pamphlet of Spiller's speculations and my estimate of them which was first published in The Popular Science Monthly, January, 1874. It is to be regretted that Isenkrahe, before publishing his essay, had not seen William B. Taylor's important memoir, hereinafter referred to, on "Kinetic Theories of Gravitation."

‡ L'unité, etc., p. 51 *seq.*

two atoms endowed solely with translation, or, again, *impinging on one another so that they can not rebound*" (which would happen if rotating atoms collided in the direction of their axes of rotation). "Evidently the atoms will remain united in the same way as the bodies called 'hard' by the mechanicians, and they will form a system animated by the movement of translation resulting from the two other movements. This system will be able to act like a single corpuscule whose mass is double, triple or generally a multiple of that of a simple atom according as two or a greater number of atoms are thus united. Here we have an obvious instance of an aggregate of atoms bound to each other, not by the influence of any sort of attraction, but by simple inertia." Judging from this passage, Secchi could hardly have been ignorant of the fact that the collision of rotating inelastic bodies does not always result in pseudo-elastic resilience. And in its application to the phenomena of gravitation his theory is plainly destructive of its own foundations. He seeks to account for gravitation upon the assumption that the density of the æthereal medium which surrounds all ponderable bodies or molecules increases from their centers outward; * and this increase of density is said to be

* This supposition is identical with that of Sir Isaac Newton, who in his letter to Boyle (Newton's Works, ed. Horsley, vol. iv, p. 385 *seq.*), speculating on the "cause of gravity" said: "I will suppose æther to consist of parts differing from one another in subtilty by infinite degrees . . . in such a manner that from the top of the air to the surface of the earth, and again from the surface of the earth to the centre thereof, the æther is insensibly finer and finer. Imagine now any body suspended in the air or lying on the earth, and the æther being by the hypothesis grosser in the pores which are in the upper parts of the body than in those which are in the lower parts, and that grosser æther being less apt to be lodged in those pores than the finer æther below, it will endeavor to get out and give way to the finer æther below, which can not be without the bodies descending to make room above for it to go into."

a consequence of the progressive conversion of rotatory into translatory motion of the æthereal particles, so that these particles are perpetually driven from the "centers of agitation" outward. "Evidently," says Secchi,* "a center of agitation, even when it is single, provided it is animated by a movement sufficiently energetic and durable, may determine the agitation of an unlimited medium, and so modify it that the density, least at its center, increases in proportion as we approach the circumference." Secchi assigns no reason why there should be a perpetual increase in the translatory motions of the æthereal particles at the expense of their rotatory motions—why the transformation should always, or generally, be from rotatory into translatory motion and not conversely; nor does he indicate the source of that "energetic and durable" agitation at the center which is said to be productive of a continual agitation of a boundless æthereal sphere; so that his explanation of the phenomena of gravitation is of very questionable validity. But, waiving this; surely, if the rotatory motion of the hard particles is gradually transformed into translatory motion, there is, by his own showing, an end to their resilience, and we are again in full presence of the unsolved problem of the reconciliation between the perpetual impact of simple, hard, and therefore inelastic atoms and the conservation of their initial energy.

The difficulty, then, appears to be inherent and insoluble. There is no method known to physical science which enables it to renounce the assumption of the perfect elasticity of the particles whereof ponderable bodies and their hypothetical imponderable envelopes are said to be composed, however clearly this assumption conflicts with one of the essential requirements of the mechanical theory.

* *L. c.*, p. 538.

CHAPTER V.

THE PROPOSITION THAT THE ELEMENTARY UNITS OF MASS ARE ABSOLUTELY INERT.

Mass and motion being mutually inconvertible, mass is absolutely inert. It can induce motion in another mass only by transferring a part or the whole of its own motion. And, inasmuch as motion can not exist by itself, but requires mass as its necessary substratum, such transference can not take place unless the masses between which it occurs are in contact. All physical action, therefore, is by impact; action at a distance is impossible; there are in nature no pulls, but only thrusts; and all force is not merely (in the language of Newton) *vis impressa*, but *vis a tergo*.

The necessity of reducing all physical action to impact has been a persistent tenet among physicists ever since the birth of modern physical science. And yet, here again, as in the cases discussed in the two preceding chapters, science rises in revolt against its own fundamental assumptions. Its first and greatest achievement was Newton's reduction of all the phenomena of celestial motion to the principle of universal gravitation—to the principle that all bodies whatever attract each other with a force proportional directly to their masses and inversely to the squares of the distances between them.

That the doctrine of universal gravitation, in the

sense of an attraction at a distance without the intervention of a medium capable of propagating mechanical impulses, is at variance with the elements of the mechanical theory was felt by no one more distinctly than by Newton himself. At the very outset of his *Principia* he carefully guarded against the imputation that he looked upon gravity as an essential and inherent attribute of matter or believed the mutual attraction of bodies to be an ultimate physical fact. The force which urges bodies in their central approach was to him, as he expressly says, a purely mathematical concept involving no consideration of real and primary physical causes.* And, evidently apprehensive lest this disclaimer should, after all, be lost sight of, he repeated it, in terms no less explicit, at the close of his great work. "The reason of these properties of gravity," he said, "I have not, as yet, been able to deduce; and I frame no hypotheses."† If, after this, there were still room for doubt as to Newton's opinions respecting the nature of gravity, it would be removed by the well-known passage in his third letter to Bentley. "It is inconceivable that inanimate brute matter should, without the mediation of something else which is not material, operate upon and affect other matter, without mutual contact, as it must do if gravitation, in the sense of Epicurus, be essential and inherent in it. And this is the reason why I desired you would not ascribe innate gravity to me. That gravity should be innate, inherent and essential to matter, so that one body may act upon

* "*Mathematicus duntaxat est hic conceptus. Nam virium causas et sedes physicas jam non expendo.*" Princ., Def. viii.

† "*Rationem vero harum gravitatis proprietatum nondum potui deducere; et hypotheses non fingo.*" Princ., Schol. Gen. ad fin. The same disclaimer is implied in the words of a scholium to the 29th Theorem, Prop. 69, Book I, of the *Principia*.

another at a distance, through a vacuum, without the mediation of anything else by and through which their action may be conveyed from one to another, is to me so great an absurdity that I believe no man, who has in philosophical matters a competent faculty of thinking, can ever fall into it. Gravity must be caused by an agent acting constantly according to certain laws; but, whether this agent be material or immaterial, I have left to the consideration of my readers." *

There is still further evidence that Newton regarded universal gravitation as a secondary phenomenon, to be explained on the principles of ordinary impact or pressure. In the later edition of his Opticks he propounds certain "Queries" relating to the possibility of deducing some of the properties of light from the undulations of an all-pervading æther, and adds (Query 21): "Is not this medium much rarer within the dense bodies of the sun, stars, planets, and comets, than in the empty celestial spaces between them? And, in passing from them to great distances, doth it not grow denser and denser perpetually, and thereby cause the gravity of those great bodies towards one another, and of their parts towards the bodies; every body endeavoring to go from the denser parts of the medium towards the rarer." †

* Newton's Works, ed. S. Horsley, vol. iv, p. 438. Zoellner (Principien einer electrodynamischen Theorie der Materie, vol. i, preface) attempts to break the force of this and other passages in the writings of Newton, but, as it appears to me, wholly without avail.

† Opticks, 4th ed., p. 325. The "Queries" appeared for the first time in the second edition of the Opticks, in the preface to which Newton again says: "To shew that I do not take gravity for an essential property of bodies, I have added one question concerning its cause, chusing to propose it by way of a question, because I am not satisfied about it for want of experiments." I have already cited in another place (*supra*, p. 42) a similar exposition of his views in the letter to Boyle.

Notwithstanding these explicit declarations, Newton's contemporaries took alarm at the apparent return of occult causes into the domain of physics. It is interesting to note the energy with which the philosophers and mathematicians of his day protested against the assumption of physical action at a distance. Huygens did not hesitate to say that "Newton's principle of attraction appeared to him absurd." Leibnitz called it "an incorporeal and inexplicable power"; John Bernoulli, who sent to the Academy of Paris two essays, in which he sought to explain the movements of the planets by an improved form of the Cartesian theory of vortices, denounced "the two suppositions of an attractive faculty and a perfect void" as "revolting to minds accustomed to receiving no principle in physics save those which are incontestable and evident." Nor did the principle of distant action find greater favor with the physicists and astronomers of a later generation. Euler observed that the action of gravity must be due either to the intervention of a spirit or to that of some subtle material medium escaping the perception of our senses; and he insisted that the latter was the only admissible alternative, although the exact demonstration of the origin of gravitative force might be difficult or impossible.* His great rival and antagonist, D'Alembert, relegated gravity to that class of causes productive of motion whose real nature is to us entirely unknown, in contradistinction to action by impact, of which we have a clear mechanical conception.† And,

* Euler, "Theoria motûs corporum solidorum," p. 68. See also his "Lettres à une princesse d'Allemagne," No. 68. October 18, 1760.

† D'Alembert, "Dynamique" (2me éd.), p. ix *seq.* It is well known how slowly and reluctantly the Newtonian philosophy found recognition and acceptance in France, where Cartesianism held undisputed sway almost to the end of the eighteenth century. What the Cartesians gener-

in spite of the assertion of John Stuart Mill and others that the thinkers of our own time have emancipated themselves from the old prejudice against *actio in distans*, it is easy to show that it is almost, if not quite, as prevalent now as it was two centuries ago. To cite but a few instances: Professor Challis, who has spent a number of years in the effort to establish a complete hydro-dynamical theory of attraction, says: "There is no other kind of force than pressure by contact of one body with another. This hypothesis is made on the principle of admitting no fundamental ideas that are not referable to sensation and experience. It is true that we see bodies obeying the influence of an external force, as when a body descends toward the earth by the action of gravity; so far as the sense of sight informs us, we do not in such cases perceive either the contact or the pressure of another body. But we have also the sense of touch or of pressure by contact—for instance, of the hand with another body—and we feel in ourselves the power of causing motion by such pressure. The consciousness of this power and the sense of touch give a distinct idea, such as all the world understands and acts upon, as to how a body may be moved; and the rule of philosophy which makes personal sensation and experience the basis of scientific knowledge, as they are the basis of the knowledge that

ally thought of the distant action of gravitation may be gathered from a paper read by Saurin to the Académie des Sciences in 1709, from which Edleston ("Correspondence between Newton and Cotes," p. 213) makes the following quotation: "Il (Newton) aime mieux considérer la pesanteur comme une qualité inhérente dans les corps et ramener les idées tant décriées de qualité occulte et d'attraction." If we abandon mechanical principles (i. e., the principles of mechanical impact and propulsion), he continues, "nous voila replongés de nouveau dans les anciennes ténèbres du peripatétisme dont le ciel nous veuille préserver."

regulates the common transactions of life, forbids recognizing any other mode than this. When, therefore, a body is caused to move without apparent contact and pressure of another body, it must still be concluded that the pressing body, although invisible, exists, unless we are prepared to admit that there are physical operations which are, and ever will be, incomprehensible by us. This admission is incompatible with the principles of the philosophy I am advocating, which assume that the information of the senses is adequate, with the aid of mathematical reasoning, to explain phenomena of all kinds. . . . All physical force being pressure, there must be a medium by which the pressure is exerted." * With equal vigor the "assumption" of universal attraction is reprobated as "an absurdity" by James Croll. "No principle" he contends, "will ever be generally received that stands in opposition to the old adage, 'A thing can not act where it is not,' any more than it would were it to stand in opposition to that other adage, 'A thing can not act before it is or when it is not.'" † Secchi protests in almost the same words. "We have said elsewhere," he declares, "how impossible it is to conceive what is called an attractive force in the strict sense of the term, that is, to imagine an active principle having its seat within the molecules and acting without a medium through an absolute void. This amounts to an admission that bodies act upon each other at a distance, that is, where they are not: an absurd hypothesis—equally absurd in the case of enormous and in that of very small distances." ‡ Friedrich

* "On the Fundamental Ideas of Matter and Force in Theoretical Physics," Phil. Mag., 4th series, vol. xxxi, p. 467.

† "On Certain Hypothetical Elements in the Theory of Gravitation," Phil. Mag., 4th series, vol. xxxiv, p. 450.

‡ L'unité, etc., p. 532 *seq.*

Mohr (who appears to be entitled to the honor of having distinctly announced the principle of the conservation of energy, even before Julius Robert Mayer) formulates his scientific creed in a number of "Theses," among which is this: "Gravity can not act except by the interposition of ponderable matter." * So also E. Du Bois-Reymond: "Forces acting through void space are in themselves inconceivable, nay absurd, and have become familiar concepts among physicists since Newton's time from a misapprehension of his doctrine and against his express warning." † And finally Balfour Stewart and P. G. Tait: "Of course, the assumption of action at a distance may be made to account for anything; but it is impossible (as Newton long ago pointed out in his celebrated letter to Bentley) for any one 'who has in philosophical matters a competent faculty of thinking' for a moment to admit the possibility of such action." ‡

The most conclusive evidence, however, of the repugnance between the assumption of distant attraction and the elementary concepts of mechanical action is found in the incessant renewal, by distinguished men since Newton's day, of the attempts to account for the phenomena of gravitation on the principles of fluid pressure or solid impact.# These attempts have recently

* "Nonnisi materiâ ponderabili interpositâ attractio agere potest." Geschichte der Erde, Appendix, p. 512.

† Ueber die Grenzen des Naturerkennens, etc., p. 11.

‡ The Unseen Universe, third ed. (1875), p. 100.

Some of these attempts are very ably discussed in a recent memoir by William B. Taylor: "Kinetic Theories of Gravitation," Smithsonian Report, 1876. This interesting essay, though quite exhaustive in the enumeration of the theories of English and French origin, might be supplemented by a collection of references to German articles and books on the same subject. See, e. g., Schramm, "Die allgemeine Bewegung und Materie," Wien, 1872; Aurel Anderssohn, "Die Mechanik der Gravita-

become reinvested with an extraordinary interest in consequence of the results of certain experiments of Professor Guthrie, who found that light bodies suspended near a vibrating disk were drawn toward it "as by an invisible chord"—a phenomenon which, as Sir William Thomson has pointed out, is explained by the fact that in a moving fluid the pressure is least where the average energy of motion is greatest.*

In the eyes of modern physics all modes of action which appear to be propagated radially from a center are progressive oscillations in elastic media. It is natural, therefore, to look for the physical cause of gravitation in the same direction. Numerous theories have been advanced in which gravitation is referred to the wave-motion of an elastic interstellar and interatomic fluid similar to, or identical with, the luminiferous æther. The most noteworthy of these theories is that

tion," Breslau, 1874 (containing a photograph of the results of an experiment in which the effects of gravitation are simulated by a ball floating in water agitated by a series of radial impulses); "Zur Loesung des Problems ueber Sitz und Wesen der Anziehung"—47 Versammlung deutscher Natureforscher und Aerzte zu Breslau, 1874; Hugo Fritsch, "Theorie der Newton'schen Gravitation und des Mariotte'schen Gesetzes," Koenigsberg, 1874; Ph. Spiller, "Die Urkraft des Weltalls," Berlin. 1876, etc. It is somewhat strange that Mr. Taylor should have omitted all reference to Huygens's elaborate "Dissertatio de causâ gravitatis" (Hugenii, Opp. Reliqua, vol. i, p. 95 *seq.*, Amstelod., 1728), as well as to the equally elaborate theory of P. A. Secchi, to which allusion has already been made in the fourth chapter. In our own country Professor Pliny Earle Chase has made large contributions to this class of literature.

* Guthrie's experiments had been anticipated, without his knowledge, by Guyot, Schellbach, and others, as appears from a communication of Guthrie himself to the Philosophical Magazine (fourth series, vol. xli, p. 405 *seq.*). Experiments similar to those of Aurel Anderssohn were made long ago by Hooke and Huygens, both of whom showed that bodies floating on water agitated by waves were drawn toward the center of agitation. Cf. Hugenii, "Diss. de causâ gravitatis," Opp. Reliqua, i, p. 99 *seq.*

of Professor Challis, who assumes that all space is filled with a vibrating æther which "is a continuous elastic medium perfectly fluid and pressing proportionally to its density." Challis, though very desirous of avoiding the cumulation of hypothetical media, and endeavoring to construe gravitative action as an incidental or residual effect of the luminar and thermal vibrations (resorting, for this purpose, to investigations analogous to those of Daniel Bernoulli, who attempted, more than a century ago, to show that the relative motions of bodies composing a material system are compounds of simple, regular, and permanent oscillations of different kinds) is constrained at last to suggest that there may be an æther of a higher order "having the same relation to the first as that has to air, and so on *ad libitum*," and that "the form of gravity is due to the attractive action of a molecule of a higher order as to magnitude than the molecule of molecular attraction." I shall have occasion, in a subsequent chapter, to discuss the scientific value of theories of this sort, in which facts are explained by an indefinite number of arbitrary assumptions multiplied in proportion to the emergencies created by the theories themselves; for the present it is sufficient to observe that all hydro-dynamical theories of gravitation are obnoxious to the fatal criticism of Arago: "If attraction is the result of the impulsion of a fluid, its action must employ a finite time in traversing the immense spaces which separate the celestial bodies,"* whereas there is now no longer any reason to doubt that the action of gravity is instantaneous. If it were otherwise—if gravity, like light or electricity, were propagated with a measurable velocity—there would necessarily be a composition of this velocity with the

* Astronomie populaire, vol. iv, p. 119.

angular orbital velocities of the planets resulting in their acceleration; the apparent line of attraction would be directed to a point in advance of the real place of the sun, just as the sun's apparent position is displaced in the direction of the earth's orbital motion by the aberration of light. Such an effect, if it had any existence, would have been detected long ago. There was a time when the action of gravity was supposed to be progressive. Daniel Bernoulli attributed the non-coincidence of the tides with the passage of the moon through the meridian to the comparative slowness of its propagation; and, at a later period, Laplace conceived for a moment that the gradual acceleration of the moon's mean motion (first ascertained by Halley by a comparison of modern lunar eclipses with those recorded by Ptolemy and the Arabian astronomers) might find an explanation in the transmission of the impulse of gravity with a velocity exceeding that of light not less than eight million times. But the retardation of the tides is now known to be a consequence of the inertia of the water and of the obstacles which it encounters in its flow; and the acceleration of the moon's motion was soon shown, by Laplace himself, to be caused, in great part at least, by the secular diminution of the eccentricity of the orbit of the earth. For this reason Laplace did not hesitate to declare that, if the action of gravity was propagated in time, its velocity must be at least fifty million times greater than that of light. It is true that the cause assigned by him for the phenomenon in question has since been found to be inadequate to its production. From a revision of the calculations of the French astronomer by Mr. Adams, some years ago, it appeared that the diminution of the eccentricity of the earth's orbit could at best account for a lunar acceleration of six seconds in one century, instead of ten seconds, the amount of acceleration assumed by Laplace, and furthermore, that the acceleration amounted in fact to nearly twelve seconds. A part of the phenomenon, therefore, had to be traced to other causes; and this has been successfully effected by showing its dependence upon the tidal retardation of the diurnal motion of the earth which occasions an apparent acceleration of the mean motion of the moon.

There is thus an entire failure of analogy, in this respect, between the action of gravity and the other known modes of physical action that are referred to æthereal undulations, such as light, radiant heat and electricity, all of which are propagated with a finite velocity. There are, moreover, as Mr. Taylor has observed, other features of gravitation which give rise to the presumption that it is of a nature essentially different from that of other forms of radial action. The action of gravity is wholly unsusceptible of interference by intervening obstacles, or, as Jevons expresses it,* "all bodies are, as it were, absolutely transparent to it;" its direction is in right lines between the centers of the attracting masses, and is not subject to reflection or refraction; unlike the forces of cohesion, capillarity, chemical affinity and electric or magnetic attraction, it is incapable of exhaustion, or rather saturation, every body attracting every other body in proportion to its mass; it is wholly independent of the nature, volume, or structure of the bodies between which it occurs, and its energy is unchangeable, incessant and inexhaustible.

On the whole, it may be safely said that the undulations of a supposed cosmical æther can not be made

* Principles of Science, vol. ii, p. 144.

available as a basis for a physical theory of gravitation, and that, if such a theory is to be framed, resort must be had to the analogies of the kinetic theory recently introduced into the science of thermo-dynamics. This is very frankly admitted by leading physicists of the day. "All attempts yet made," say Stewart and Tait,* "to connect gravitation with the luminiferous æther, or the medium required to explain electric and magnetic distance-action, have completely failed, so that we are apparently driven to the impact theory as the only possible one." The only impact theory seriously discussed by modern physicists and astronomers is that of Le Sage,† which, stated in a few words, is this: Space is constantly traversed in all directions by streams of infinitely small bodies moving with an almost infinite velocity and coming from unknown regions of the universe. These bodies are termed "ultramundane corpuscules." By reason of their minuteness they rarely, if ever, collide, and the greater part of them find ready passage through ordinary sensible bodies, so that all parts of these bodies—those in the interior as well as those on the surface—are equally liable to be struck by the corpuscules, the force of the impact being thus proportional, not to the surfaces, but to the masses of the bodies. A single body or particle would be equally battered by these corpus-

* The Unseen Universe, § 140.

† Arago suggests (Astr. pop., iv, p. 118) that the theory of Le Sage is simply a reproduction, in an improved form, of the systematic ideas of Fatio de Duillers (the insane and meddlesome partisan of Newton in his controversy with Leibnitz respecting the priority in the invention of the differential calculus) and Varignon, which had been communicated to Le Sage before their publication. But this is probably an error; Varignon's speculations, at least, were similar to those of Newton in the 21st Query of his "Opticks."

cules on all sides; but any two bodies act as mutual screens, so that each receives a less number of impacts on the side facing the other. They are consequently driven toward each other. The motion of the corpuscules being rectilinear in all directions, the diminution of pressure thus resulting is inversely as the squares of the distances between the bodies affected.

With all deference due to the authority of the scientific men by whom this theory has been countenanced, it must be said that the extravagance of its assumptions at once characterizes it as a survival of the fancies of an age in which the functions of a scientific theory were imperfectly understood. Its intellectual consanguinity with the old vortices and harmonic circulations is unmistakable. It utterly ignores the necessity of accounting for the origin of the enormous energy constantly expended by the supposed streams of ultramundane corpuscules; both the agency postulated and the mode of its action are unknown to experience; and it is doubtful whether its assumptions, if they could be granted, would serve as an explanation of all or any of the features of gravitation in the presence of which, as we have seen, every hydro-dynamic theory is doomed to failure. The futility of Le Sage's theory, however, is most strikingly exhibited by Clerk Maxwell,* who tests it by the principle of the conservation of energy. If the ultramundane corpuscules impinging upon sensible bodies are perfectly elastic and rebound with the same velocity with which they approach, they will "carry their energy with them into the ultramundane regions," and in that event "the corpuscules rebounding from the body in any given direction will be both in number and velocity exactly equivalent to those which are prevented from

* Encyclopædia Britannica, s. v. "Atom."

proceeding in that direction by being deflected by that body, whatever the shape of the body and however many bodies are in the field." In this case, therefore, there is no gravitative action. If, on the other hand, the corpuscules are inelastic, or imperfectly elastic—inasmuch as the action of gravity is supposed to be due to the comparatively small difference between the impacts on opposite sides of the body—the energy of those impacts, at least, which balance each other, must be (partially or wholly, according to the degree of corpuscular elasticity) converted into heat, and "the amount of heat so generated would in a few seconds raise the body, and in like manner the whole material universe, to a white heat."*

Once more, then, science is in irreconcilable conflict with one of the fundamental postulates of the mechanical theory. Action at a distance, the impossibility of which the theory is constrained to assert, proves to be an ultimate fact inexplicable on the principles of impact and pressure of bodies in immediate contact. And this fact is the foundation of the most magnificent theoretical structure which science has ever erected—a foundation deepening with every new reach of our telescopic vision, and broadening with every further stretch of mathematical analysis.

* Mr. S. Tolver Preston has recently (Phil. Mag., September and November, 1877, and February and May, 1878) proposed a modification of Le Sage's theory, in which he seeks to dispense with the ultramundane feature of the corpuscules, and to account for gravitation upon the postulate of the kinetic theory of gases alone. His theory is founded on the assumptions that "the range of gravity is limited," and that "the stars move in straight lines and not in orbits." In view of these assumptions, and of my discussion of the kinetic theory of gases in a separate chapter, I do not deem it necessary to devote any space to it here.

CHAPTER VI.

THE PROPOSITION THAT ALL POTENTIAL ENERGY IS IN REALITY KINETIC.—EVOLUTION OF THE DOCTRINE OF THE CONSERVATION OF ENERGY.

According to the mechanical theory, motion, like mass, is indestructible and unchangeable; it can not vanish and reappear. Any change in its rate results from its distribution among a greater or less number of units of mass. And, motion and mass being mutually inconvertible, nothing but motion can be the cause of motion. There is, therefore, no potential energy; all energy is in reality kinetic.

The close logical connection of this proposition with that discussed in the last chapter is obvious, and has not escaped the notice of leading physicists. Stewart and Tait, after giving an account of Le Sage's hypothesis, which, in their opinion, contains the rudiments, at least, of the only tenable physical theory of gravitation, proceed to say: "If Le Sage's theory, or anything of a similar nature, be at all a representation of the mechanism of gravitation, a fatal blow is dealt to the notion of the tranquil form of power we have called potential energy. Not that there will cease to be a profound difference in kind between it and ordinary kinetic energy, *but that* BOTH *will be henceforth to be regarded as kinetic.*" * This declaration has re-

* The Unseen Universe, § 142.

cently been repeated by Professor Tait in his lecture on Force.*

The proposition here insisted upon is irrecusable by any consistent advocate of the mechanical theory. But, again, modern science peremptorily refuses its assent. It asserts that all, or nearly all, physical changes in the universe are mutual conversions of kinetic and potential energies—that energy is incessantly stored as virtual power and restored as actual motion. When the bob of an ordinary pendulum descends from its highest to its lowest point, its potential energy diminishes in proportion to the increase of its actual motion; when it rises again, its energy of motion disappears at the same rate up to its arrival at its highest point opposite the first, where it is for an instant motionless, all its energy being due to its position. And this conversion and reconversion of the two forms of energy are typical alike of the supposed oscillations of the ultimate atoms or molecules and of the orbital swing of the large bodies composing a planetary system. A planet moving in an excentric orbit gains energy of motion as it approaches the sun and loses it again in the same proportion as it recedes from it. The same mutual transformation is exhibited in another wide domain of physical phenomena: action due to chemical affinity. A lump of coal lies buried in the earth for a million years; during all this time there is no appreciable change in its position as referred to surrounding objects, or in the relative positions of its parts—it is without external or internal motion (except that which it shares with the planet of which it is a part); now we bring it to the surface, into the atmosphere containing oxygen and into contact with

* On some Recent Advances in Physical Science, second ed., pp. 262, 263

a flame; its latent power at once becomes sensible—it burns, giving rise to vigorous action which manifests itself as light and heat. The tendency of modern science is to trace all physical change to a few primary forms of potential energy, chief among which are gravity and chemical affinity. In the opinion of modern physicists, the only plausible theory, thus far advanced, of the origin of stellar and planetary systems is that known as the nebular hypothesis; and, whether we adopt its familiar Kant-Laplacean form, or one of its more recent modifications, in either case all the molar, if not the molecular, forces of the universe are ultimately derived from the attraction due to the mere position of the original particles supposed to be uniformly diffused in space. And all changes in the comparatively minute organic or inorganic forms are referred, proximately at least, in physiology as well as in physics, to the affinities of the chemical elements.

In truth, modern science teaches that diversity and change in the phenomena of nature are possible only on condition that energy of motion is capable of being stored as energy of position. The relatively permanent concretion of material forms, chemical action and reaction, crystallization, the evolution of vegetal and animal organisms—all depend upon the "locking up" of kinetic action in the form of latent energy. To make this clear, and to show that the effort to abolish the distinction between kinetic and potential energy is without avail, it will be useful briefly to review the history of the doctrine of the conservation of energy.

In a general sense, this doctrine is coeval with the dawn of human intelligence. It is nothing more than an application of the simple principle that nothing can come from or to nothing.* But the history of its development and adhibition in physical science begins with its emphatic statement in the "Principia Philosophiae" of the inventor of the system of cosmical vortices.†

* It may be truly asserted that human intelligence begins and ends with the principle above stated. When all the phenomenal changes in the universe shall have been reduced to the one principle of the conservation of energy, the time will have come for celebrating the final consummation of physical science in a new epic "*de rerum naturâ;*" and in its first chapter will again be written the words of Lucretius:

"... *res* ... *non posse creari*
De nihilo, neque item genitas in nil revocari."

It is not a little curious to note the unaminity and emphasis with which the early Greek philosophers gave utterance to the declaration that nothing could absolutely originate or perish—the rudimentary form of the law of causality. Diogenes of Apollonia declared: "*οὖδεν ἐκ τοῦ μη ὄντος γίνεσθαι*" (Diog. Laert., ix, 57); Parmenides: "*ὡς αγένετον ἐὸν και ἀνώλεθρον ἐστιν*" (Karsten, Rel., v, 58); Empedocles: "*ἐκ τοῦ γὰρ μη ἐόντος ἀμήχανον ἐστὶ γενέσθαι*" (Karsten, v, 48); Democritos: "*μηδέν τ' ἐκ τοῦ μὴ ὄντος γίνεσθαι καὶ εἰς τὸ μὴ ὂν φθείρεσθαι*" (Diog. Laert., ix, 44). The first application of this principle to motion was made by Epicurus (Diog. Laert., lib. x; Lucret. "De rer. nat.," vv. 294–307), who sought to demonstrate the conservation of both mass and motion by the argument that there is no place beyond the universe to which matter or motion could be communicated or from which it could be derived—an argument which was reproduced by Leibnitz (Opp. Math., vol. vi, p. 440—cf. Berthold, "Notizen," etc., in Pogg. Ann., vol. clvii, p. 342), and which is in effect a shrewd anticipation of the modern concept of a "conservative system." An elaborate exposition of the Epicurean doctrine is given by Gassendi ("Ad librum decimum Diogenis Laertii Notae," opp., ed. Lugd., vol. iii, p. 241 *seq.*). It is not improbable that this exposition had its influence on the meditations of Descartes, notwithstanding the wide divergence between his philosophical tendencies and those of Gassendi.

† Descartes has been called the father of modern philosophy; with equal propriety he might also be designated as the father of modern physical science. His title to the honors of paternity in philosophy, no less than in physics, must find other muniments than the discovery, or even exact statement, of permanently valuable truths. Few of his philosophical tenets endure, at least in the form in which he held them, and some of the truths which he rejected are now counted among our most indispen-

Descartes announced the doctrine of the conservation of motion in terms perfectly explicit. He declared that God was the spring-head of motion, and always conserved in the world the same quantity of motion.*

sable possessions. As a physicist he broached a number of theories that have proved to be wholly unfounded, and he ignored or misconceived almost all the laws of mechanical action whose discovery constituted the distinction of his older contemporary, Galilei. In philosophy he was the immediate progenitor of Spinoza, whose system, though in effect an involuntary *reductio ad absurdum* of all ontological speculation, has served, by reason of the specious elegance of its pseudo-mathematical paralogisms, to retard the discovery of true principles of philosophical inquiry to an incalculable extent. In physics his vagaries obscured the field of investigation to such a degree that the shadows have not wholly vanished to this day. Though professing to emancipate himself from the metaphysical traditions of the period which was then near its close, he was thoroughly imbued with their spirit. But precisely for this reason his writings influenced the thought of the seventeenth century more extensively than the researches of those who resorted to the scientific methods of experiment and observation—methods that were wholly at variance with the mental habits of the age. He was essentially a metaphysician, an ontologist of the mediæval type; but he discussed nearly all the problems whose solution was the task devolving upon the physicists and mathematicians of the two centuries that have elapsed since his day. Thus his speculations, though on the whole nugatory in themselves, became the ferment which induced the process of gradual clarification in the rapidly thickening mixture of scientific material. This ferment was not the less important because it was almost wholly lost in the progress of its action.

In saying all this I have no disposition to detract from the general admiration due to the vigor and acuteness of his intellect; nor do I forget that he is the founder of analytical geometry. And it is not necessary, I trust, to add that, while I give candid expression to my estimate of the value of Spinoza's philosophical system, I am not a stranger to the emotion which will always be felt when the touching figure of the lonely thinker rises into view, and that I am not insensible to the charm of the simple beauty of a life which, more perfectly, perhaps, than any other, exemplifies the Tusculan definition: *vivere est cogitare.*

* "Generalem (motûs causam) quod attinet, manifestum mihi videtur illam non aliam esse quam Deum ipsum qui materiam simul cum motu et quiete in principio creavit, jamque per solum suum concursum ordina-

If he had not been precluded (by his assumption that the only primary properties of matter were extension and mobility) from admitting the atomic constitution of matter, he would, no doubt, have asserted the conservation of motion in the sense which is generally attributed to the principle of the conservation of energy in our day by persons without scientific training: that the atoms of which the material world is composed are perpetually in a state of uniform translatory or oscillatary motion, changing only in direction, or, if they move with different velocities, that the sum of these velocities is constant. In view of his general physical theory, Descartes was constrained to resort, not to the atom—the supposed primordial unit of mass, the existence of which he denied—but to mass generally; and the conservation of motion in his system assumed the form of a conservation of the quantity of motion in the sense of the sum of the products of all masses into their respective velocities.* It is worthy of note that the term "quantity of motion" as expressive of the product of a mass into its velocity (i. e., momentum) was adopted by Newton, and has maintained itself in physics to the present day.

It is manifest that the conservation of motion, as an

rium, tantundem motûs et quietis in eâ totâ, quantum tunc posuit, conservat." Princ. Phil., ii, § 36. The doctrine is stated, substantially in the same terms, in various other parts of the same work, e. g., ii, § 42; iii, § 46.

* The vagueness of Descartes's mechanical notions is strikingly exhibited in his efforts to reconcile this with his third law of motion, according to which a body loses no motion in a collision with a "stronger" one—"ubi corpus quod movetur alteri occurrit, si minorem habeat vim ad pergendum secundum lineam rectam, quam hoc alterum ad ei resistendum, et motum suum retinendo solam motûs determinationem amittit; si vero habeat majorem, tunc alterum corpus secum movet ac quantum ei dat de suo motu, tantundem perdit." Princ. Phil., ii, § 40.

absolute quantity in the popular sense (in which it is, in fact, a conservation of velocities), would be possible only in a world without differences of density or structure. If motion were conserved in this sense, there could be neither phenomenal diversity, nor phenomenal change. To the universe as we know it, with its incessant transformations, the assumed principle of the conservation of motion can have no application. This was seen, dimly at least, by Leibnitz, who denied that there was any conservation of motion in the Cartesian sense. His denial found its most pointed expression in an essay entitled "Short demonstration of the memorable error of Descartes and others in regard to a law of nature, according to which, as they claim, God always conserves the same quantity of motion, which they also abuse in mechanics." * To the Cartesian doctrine of the conservation of the quantity of motion he opposed the principle of the conservation of *vis viva*—of the product of mass into the square of its velocity.

Here was the origin of the famous controversy between the Leibnitians and Cartesians, respecting the true measure of the forces in the universe, which was participated in by so many mathematicians and philosophers, and to which, as is well known, a late and inapposite contribution was made by Kant. This controversy has long since been finally settled; but it is so important for my ulterior purpose to clear up the prevalent misconceptions of the true import of the principle of the conservation of energy, that I devote a moment's consideration to its merits.

* "*Brevis demonstratio erroris memorabilis Cartesii et aliorum circa legem naturae, secundum quam volunt a Deo eandem semper quantitatem motûs conservari, quâ et in re mechanicâ abutuntur.*" Acta Erud., Lips., 1686 (Leibn., opp. math., vol. vi, p. 117).

Force in its ordinary sense (as the cause of motion, or rather, as the aggregate of all its conditions) finds its measure simply in the velocity of a unit of mass. Thus force and mass are measured by each other. Two forces are the same when they generate the same velocity (or, more generally, the same acceleration) in the same mass; and two masses are the same when they are equally accelerated by the same force. When the motion of a unit of mass is distributed among several units, the motion of each unit becomes less in proportion to the number of units among which the distribution is made. The velocity (or acceleration) of a body is therefore directly as the force, and inversely as the mass. And, in the case of constant forces producing uniform accelerations, the velocities are obviously proportional to the times of action.

We have, therefore,

$$\text{Velocity} = \frac{\text{Force}}{\text{Mass}} \times \text{Time of Action, or,}$$

$$\textit{Mass} \times \textit{Velocity} = \textit{Force} \times \textit{Time of Action}\,; \quad . \;.\; . \quad (1)$$

i. e., the force exerted during any given time is equal to the product of the mass into the velocity. On the other hand, the space or distance through which a body moves under the action of a constant force is, like velocity, directly as the force and inversely as the mass; but, unlike velocity, it is proportional, not to the time simply, but to half the square of the time of action. Hence,

$$\text{Space or distance of Action} = \frac{\text{Force}}{\text{Mass}} \times \tfrac{1}{2}\,(\text{Time of Action})^2,$$

or (inasmuch as, according to the first equation,

$$\text{Time of Action} = \frac{\text{Mass} \times \text{Velocity}}{\text{Force}}\Big)$$

$$\tfrac{1}{2}\,\textit{Mass} \times \textit{Velocity}^2 = \textit{Force} \times \textit{Distance of Action} \ldots (2).$$

The first term of this last equation—the product of the mass into half the square of the velocity—is the Leibnitian *vis viva*, and is now termed kinetic energy.*

It is apparent that the first (Cartesian) formula indicates the measure of a given force *during a given time of action*, while the second (Leibnitian) formula contains the measure of the force *acting through a given distance.* There is no inconsistency between the two; on the contrary, the one is a corollary from the other. And yet the controversy is of interest in view of the Cartesian claim (which survives as an indelible fancy in many minds) that force, in the sense of the rate of the generation or transference of quantity of motion, is conserved, and that the momenta during any two equal intervals of time are the same. In the light of modern science nothing is more demonstrably untrue than the doctrine of the conservation of motion as it was held by Descartes. Nevertheless, there is a sense in which the quantity of motion—or what is now usually called momentum—is constant in the mutual actions of bodies composing a material system. Momentum being the product of mass into the velocity, and velocity being necessarily in a definite direction, it follows, as Newton himself has shown, from his third law (according to which action and reaction are equal and opposite—all force, so called, being but one aspect of the mutual equal and opposite action of two bodies—) that the momentum of any system of bodies, i. e., the sum of their quantities of motion, in whatever direction these quantities be measured, is never changed by their mut-

* Leibnitz and his contemporaries designated the *whole* product of the mass into the square of the velocity as the *vis viva*; but this is correct only when the measure of forces is stated in the form of a proportion.

ual action. Whatever momentum is acquired by any part of the system is lost by another part in the same direction. From this follows the important dynamical principle (announced in Newton's fourth corollary from his laws of motion) that the center of inertia of a system of bodies is never affected by their mutual action.

To interpret the Cartesian proposition in its application to the universe as a single conservative system, so as to make it conformable to fact, it would be necessary to take some one fixed direction and project upon it all the motions of its constituent bodies or particles —in other words, to take their effective components as represented by the cosines of the angles between their several directions and the standard direction to which they are referred. This being done, the sum of the momenta, i. e., of the products of all the masses into their velocities in the direction indicated, would be constant; it being understood that, if motion in one direction is taken as positive, motion in the opposite direction (and hence also the momentum whereof it is a factor) is negative.*

Although the merit of having formulated the principle of the conservation of *vis viva* belongs to Leibnitz, the first clear statement of the relation of this principle to that of the conservation of momentum is

* It is sometimes said that quantities of motion partially or wholly neutralize or destroy each other, as in the case of the central collision of two inelastic bodies moving with equal velocities in diametrically opposite directions, where the bodies, after impact, are at rest, the resultant momentum being = 0. But the momenta of the two bodies being equal and opposite, and their sum, therefore, being that of two equal quantities, one of which is positive and the other negative, this sum was also = 0 before collision, so that the case stated is no exception to the rule that the momenta of colliding bodies are unaltered by their mutual impact.

due to Huygens, and is in these words: "The quantity of motion possessed by two bodies may be augmented or diminished by their encounter; but there remains always the same quantity on the same side, if we subtract the quantity of opposite motion. . . . The sum of the products of every hard body multiplied by the square of its velocity is always the same before and after the encounter." *

The progress made up to this point, in the rectification of the Cartesian doctrine, consisted in the denial of the conservation of motion in the sense of mere velocity or of the quantity of motion and the rate of its change irrespective of its direction, and in the assertion of the conservation of *energy* of motion—a quantity proportional to the product of mass into the square of its velocity. Such was the state of the doctrine in Newton's time.

The Leibnitian principle might, even at this time, (all the premisses being given in Newton's laws of motion, and especially in his interpretation of the third law) have been generalized so as to embrace, or to im-

* "La quantité du mouvement qu'ont deux corps se peut augmenter ou diminuer par leur rencontre; mais il y reste toujours la même quantité vers le même côté, en soustrayant la quantité du mouvement contraire . . . . La somme des produits faits de la grandeur de chaque corps dur multiplié par le quarré de sa vitesse, est toujours la même devant et après la rencontre." Cf. Akin, "On the History of Force," Phil. Mag., 4th series, vol. xxviii, p. 472. Professor Bohn (*ib.*, p. 313) claims the honor of priority in giving a clear exposition of the principle of the conservation of *vis viva* for John Bernoulli; but upon perusal of the passages quoted by him it will be seen that Bernoulli's conception rested upon the metaphysical assumption of the substantiality of motion and the equality of cause and effect. Indeed, John Bernoulli had adopted the principle in the form and upon the considerations presented by Leibnitz, who, like Descartes, was a metaphysician rather than a physicist, while Huygens, a true man of science, arrived at his propositions by a series of generalizations of special cases.

ply, not only the conservation of *vis viva*, but also the principle of virtual velocities, the conservation of momentum (including angular momentum) and the modern principle of the conservation of energy. The formula would have been this: Neither the momentum, nor the energy, of a system of bodies is ever changed by their mutual actions. It is manifest that this is nothing more than an extension of the principle of inertia according to which a body, whether it be regarded as simple or as composed of parts, can not move itself, i. e., can not produce any change in its own state of rest or of uniform motion as a whole.

Modern science has framed a number of concepts which serve to facilitate the apprehension of the laws regulating changes in the condition of material aggregates. Treating every sensible body as a system of units of mass, it defines "work" as a change in the configuration of such a system in opposition to the forces resisting it, and "energy" as the capacity to do work. Whenever such a system is considered as being under the exclusive control of the mutual forces of its constituent units, i. e., when it is neither acted upon by other systems, nor acts upon them, it is called a "conservative system." In fact there is no limited material system which is not involved in mutual actions with systems or bodies without, and for this reason a "conservative system" is more appropriately defined as a group of bodies which, in passing through any cycle of changes of configuration, does the same quantity of external work which is done upon it, so that the energy derived from bodies without is compensated for by an equal amount of energy communicated to external bodies. If, now, we express the principle of the conservation of *vis viva* in terms of these concepts, it assumes the following form: In any series of changes in the configuration of a conservative system, its actual energy (energy of motion, or *vis viva*—now termed *kinetic energy*) is the same whenever the configuration is the same, i. e., whenever its constituent units are in the same relative positions, through whatever orbits and with whatever velocities they may have moved in the passage from one configuration to the other. The import of this proposition will be best realized by considering the simple case of the oscillations of a pendulum which, ever since the days of Galileo, has served as a paradigm for the illustration of dynamical laws. The bob of the pendulum changes velocity at every point; but the velocities at points equidistant from its point of maximum velocity are equal.* A still simpler case is that of a body projected perpendicularly upward and returning to the point from which it was projected; in its ascent it is retarded, and in its descent accelerated (leaving out of account the resistance of the air), by the constant action of gravity; but at the same points the velocities of ascent and descent are the same. A similar (at bottom the same) instance is afforded by celestial bodies, moving in elliptical orbits, which—again abstracting from causes that interfere with the strict periodicity of their motions—have the same energies of motion at the same, or symmetrically corresponding, points of their orbits. The instances here adduced are all cases of varying (uniformly accelerated or retarded) motion; when the motion is uniform, the law of conservation is simply the well-known principle of virtual velocities.

Obviously the next question in order is: What is the

* This is, of course, strictly true only of an ideal pendulum, swinging *in vacuo* and without friction.

law of energy without regard to the completion of the cycle of configurative changes—in the interval, during the passage of the system from any assumed initial configuration to a different one, and during its return from this to the initial configuration? The answer to this question, which has taken definite form in very recent times, constitutes the true and exhaustive statement of the doctrine of the conservation of energy. It is this: In any series of changes in the configuration of a conservative system, the sum of its kinetic and potential energies (i. e., the actual energy of the system at a given instant added to the work done in passing from the initial configuration to the configuration at that instant) is constant—the work done being stored as power to reproduce the initial configuration and thus to restore the actual energy lost. Literally, this statement of the principle applies only to cases where work is done against the forces of the system, as, for instance, when a body is projected upward against gravity—when, therefore, kinetic energy is stored as potential energy. Whenever, conversely, kinetic energy is restored and potential energy lost, as in the case of a falling body, the statement must be so modified as to assert the constancy of the sum obtained by adding the kinetic energy due to a given configuration to the work *to be done* in reproducing the initial configuration where the potential energy is at its maximum. In such cases the mathematical expression for the potential energy in terms of work is negative. In its application to the energy of the universe (which is necessarily conservative, there being no bodies without it) the law of conservation is this: The kinetic energy of the universe *plus* the work to be done by the mutual forces of its constituent elements by removing them to the limit of exhaustion of the action of these forces, i. e., to infinite distances from each other, is at all times constant.*

The conformity of the principle of the conservation of energy to the facts of experience is sufficiently apparent whenever we deal with visible or otherwise perceptible changes in the position or configuration of a body or system of bodies, such as the action of gravity, the strain of an elastic body, etc. In these cases we readily see that energy is alternately stored as energy of position and restored as energy of motion. But there is a class of cases in which there is loss of energy of motion without manifest change of position. When two equal inelastic bodies, moving with equal velocities in opposite directions, collide centrally, there is, apparently at least, a total destruction of motion, and there is no gain of position, for the bodies remain at rest at the point of collision. A similar loss of actual energy is observed whenever work is done against friction. What becomes of the energy of motion which seems to disappear in cases of this kind? To this question Newton clearly had no definite answer. He expressly asserted that "motion may be got or lost," and that, "the *vis inertiæ* being a passive principle, . . . some other principle was necessary for putting bodies into motion, and, now they are in motion, some other principle is necessary for conserving the motion. . . . By reason of the tenacity of fluids, and attrition

* It is to be observed that I am here stating the doctrine of the conservation of energy in its application to the universe as it is generally held among physicists. The discussion of the question respecting the admissibility of applying logical concepts and mathematical formulæ based upon the conditions of finite existence to the Infinite, of dealing with the boundless world as with a definite mechanical system, and with its energy as with a constant quantity, must be reserved for a later stage in the progress of our inquiry.

of their parts, and the weakness of elasticity in solids, motion is much more apt to be lost than got, and is always upon the decay." * But it is an error to maintain, with Stewart and Tait,† that the answer was unknown in Newton's time. The answer of modern science, which is that the apparent loss of molar motion results from its real conversion into molecular motion, was anticipated by Leibnitz, as is shown in the following remarkable passage found in his fifth letter to Clarke: "I had maintained that the active forces are conserved in the world. It is objected that two soft or inelastic bodies, when they collide, lose part of their force. I answer, they do not. It is true that the 'wholes' lose it in reference to their total movement; but it is received by the particles, they being agitated inwardly by the force of the collision. Thus the loss ensues only in appearance. The forces are not destroyed, but dissipated among the minute parts. This is not losing them, but it is doing what those do who turn money into small change." ‡ The truth thus an-

* "Opticks," 4th ed., p. 373.

† The Unseen Universe, § 100.

‡ "J'avais soutenu que les *Forces actives* se conservent dans le monde. On m'objecte, que deux corps mous, ou non-élastiques, concourant entre eux, perdent de leur *force*. Je réponds que non. Il est vrai que les Touts la perdent par rapport à leur mouvement total; mais les parties la reçoivent, étant agitées intérieurement par la force du concours. Ainsi ce défaut n'arrive qu'en apparence. Les forces ne sont détruites, mais dissipées parmi les parties menues. Ce n'est pas les perdre, mais c'est faire comme font ceux qui changent la grosse monnaie en petite." Opp. phil., ed. Erdmann, p. 775. It is strange that this passage should have remained unnoticed for many years even after the adoption of the modern theory of the conservation and transformation of energy and of the correlation of forces. I found it many years ago; Du Bois-Reymond has recently called attention to it in a lecture, "Leibnizische Gedanken in der neueren Naturwissenschaft." There is another passage of the same import in Leibnitz's Mathematical Works (ed. Gerhardt), vol.

nounced was a "bedridden truth" (to use an expression of Coleridge) for a long time; in spite of the vigorous, and even violent, disputes about forces and their measure, and in the midst of the rapid accumulation of physical facts and theories, it remained barren for more than a century. This seemingly anomalous fact is explained by the circumstance that, up to the middle of the present century, heat, electricity, magnetism, etc., were supposed to be material substances whose interconvertibility with mechanical motion or energy appeared to be utterly inconceivable. It was only after the establishment of the dynamic theories of the "imponderables" that the doctrine of the conservation and transformation of energy became fertile, and led to a fundamental reconstitution of the entire body of physics.*

The correlation and mutual conversion of the various forms of energy have been so extensively illustrated in the scientific writings of the day, that it is unnecessary to dwell upon them here. The purpose of my hurried glance at the history of the doctrine of the conservation of energy, or rather, of the evolution of the scientific concepts embraced in it, was simply to show that this history is in effect that of a progressive abandonment of the mechanical proposition placed at the head of the present chapter, which is substantially identical with Descartes's theory of the conservation of mo-

ii, p. 230. Dr. Berthold has shown (Pogg. Ann., vol. clvii, p. 350) that the "allotropy of force" was announced, more than a century ago, in terms of curious precision, by Diderot in his "Pensées sur l'interprétation de la nature," Londres, 1754, § 45.

* I am aware, of course, of the anticipations of the modern theory of heat by Bacon, Locke, Rumford, Sir Humphry Davy, etc.; but their announcement, however clear, that heat is but a "mode of motion," received as little attention from contemporary physicists as the Leibnitian doctrine above referred to.

tion—a circumstance whose significance I hope to point out hereafter.

We have now discussed the four cardinal propositions of the atomo-mechanical theory, and have found (without entering upon the domain of the organic sciences) that they are severally denied by the sciences of chemistry, physics, and astronomy. Before we proceed to investigate the causes and consequences of this result and to consider the relation of the mechanical theory to the laws of thought and the history of its evolution, it is important to supplement this discussion by an inquiry into the nature, validity, and scientific value of the hypothesis of the atomic constitution of matter.

CHAPTER VII.

THE THEORY OF THE ATOMIC CONSTITUTION OF MATTER.

THE doctrine that an exhaustive analysis of matter into its real elements, if it could be practically effected, would yield an aggregate of indivisible and indestructible particles, is one of the earliest products of human speculation, and has held its ground more persistently than any other tenet of science or philosophy. It is true that the atomic theory, since its first promulgation by the old Greek philosophers, and its elaborate statement by Lucretius, has been modified and refined. There is probably no one, at this day, who invests the atoms with hooks and loops, or accounts for the bitter taste of wormwood by the raggedness, and for the sweetness of honey and milk by the smooth roundness of the constituent atoms.* But the atoms of modern science are still of determinate weight, if not of definite, uniform and constant figure, and stand for something more than abstract units even in the view of those who, like Boscovich, Faraday, Ampère, or Fechner, profess to regard them as mere centers of force. And there is no difficulty in stating the atomic doctrine in terms applicable alike to all the acceptations in which it is now held by scientific men. Whatever diversity of opinion may prevail as to the form, size, etc., of the atoms, all who advance the atomic hypothesis, in any of

* Lucretius, De Rerum Nat., ii, 398 *seq.*

its varieties, as a physical theory, agree in three propositions, which may be stated as follows:

1. *Atoms are absolutely simple, unchangeable, indestructible; they are physically, if not mathematically, indivisible.*

2. *Matter consists of discrete parts, the constituent atoms being separated by void interstitial spaces. In contrast to the continuity of space stands the discontinuity of matter. The expansion of a body is simply an increase, its contraction a lessening, of the spatial intervals between the atoms.*

3. *The atoms composing the different chemical elements are of determinate specific weights, corresponding to their equivalents of combination.**

Confessedly the atomic theory is but an hypothesis. This in itself is not decisive against its value; all physical theories properly so called are hypotheses whose eventual recognition as truths depends upon their consistency with themselves, upon their agreement with the canons of logic, upon their congruence with the facts which they serve to connect and explain, upon their conformity with the ascertained order of Nature, upon the extent to which they approve themselves as trustworthy anticipations or previsions of facts verified by subsequent observation or experiment, and finally upon their simplicity, or rather their reducing power. The merits of the atomic theory, too, are to be determined by seeing whether or not it satisfactorily and simply accounts for the phenomena as the explanation of which it is propounded, and whether or not it is in

* To avoid confusion, I purposely ignore, for the moment, the distinction between *molecules* as the ultimate products of the physical division of matter, and *atoms* as the ultimate products of its chemical decomposition, preferring to use the word *atoms* in the sense of the least particles into which bodies are divisible by any means.

accord with itself and with the known laws of Reason and of Nature.

For what facts, then, is the atomic hypothesis meant to account, and to what degree is the account it offers satisfactory?

It is claimed that the first of the three propositions above enumerated (the proposition which asserts the persistent integrity of atoms, or their unchangeability both in weight and volume) accounts for the indestructibility and impenetrability of matter; that the second of these propositions (relating to the discontinuity of matter) is an indispensable postulate for the explanation of certain physical phenomena, such as the dispersion and polarization of light; and that the third proposition (according to which the atoms composing the chemical elements are of determinate specific gravities) is the necessary general expression of the laws of definite constitution, equivalent proportion, and multiple combination, in chemistry.

In discussing these claims, it is important, first, to verify the facts and to reduce the statements of these facts to exact expression, and then to see how far they are fused by the theory.

1. The indestructibility of matter is an unquestionable truth. But in what sense, and upon what grounds, is this indestructibility predicated of matter? The unanimous answer of the atomists is: Experience teaches that all the changes to which matter is subject are but variations of form, and that amid these variations there is an unvarying constant—the mass or quantity of matter. The constancy of the mass is attested by the balance, which shows that neither fusion nor sublimation, neither generation nor corruption, can add to or detract from the weight of a body subjected to experiment.

When a pound of carbon is burned, the balance demonstrates the continuing existence of this pound in the carbonic acid, which is the product of combustion, and from which the original weight of carbon may be recovered. The quantity of matter is measured by its weight, and this weight is unchangeable.

Such is the fact, familiar to every one, and its interpretation equally familiar. To test the correctness of this interpretation, we may be permitted slightly to vary the method of verifying it. Instead of burning the pound of carbon, let us simply carry it to the summit of a mountain, or remove it to a lower latitude; is its weight still the same? Relatively it is; it will still balance the original counterpoise. But the "absolute weight" is no longer the same. This appears at once, if we give to the balance another form, taking a pendulum instead of a pair of scales. The pendulum on the mountain or near the equator swings more slowly than at the foot of the mountain or near the pole, for the reason that it has become specifically lighter by being farther removed from the center of the earth's attraction, in conformity to the law that the attractions of bodies vary inversely as the squares of their distances.

It is thus evident that the constancy, upon the observation of which the assertion of the indestructibility of matter is based, is simply the constancy of a relation, and that the ordinary statement of the fact is crude and inadequate. Indeed, while it is true that the weight of a body is a measure of its mass, this is but a single case of the more general fact that the masses of bodies are inversely as the velocities imparted to them by the action of the same force, or, more generally still, inversely as the accelerations produced in them by the same force. In the case of gravity, the forces of attraction are directly proportional to the masses, so that the action of these forces (weight) is the simplest measure of the relation between any two masses as such; but, in any inquiry relating to the validity of the atomic theory, it is necessary to bear in mind that this weight is not the equivalent, or rather presentation, of an absolute substantive entity in one of the bodies (the body weighed), but the mere expression of a relation between two bodies mutually attracting each other. And it is further necessary to remember that this weight may be indefinitely reduced, without any diminution in the mass of the body weighed, by a mere change of its position in reference to the body between which and the body weighed the relation subsists.

Masses find their true and only measure in the action of forces, and the persistence of the effect of this action is the simple and accurate expression of the fact which is ordinarily described as the indestructibility of matter. It is obvious that this persistence is in no sense explained or accounted for by the atomic hypothesis. It may be that such persistence is an attribute of the minute, insensible particles which are supposed to constitute matter, as well as of sensible masses; but, surely, the hypothetical recurrence of a fact in the atom is no explanation of the actual occurrence of the same fact in the conglomerate mass. Whatever mystery is involved in the phenomenon is as great in the case of the atom as in that of a solar or planetary sphere. Breaking a magnet into fragments, and showing that each fragment is endowed with the magnetic polarity of the integer magnet, is no explanation of the phenomenon of magnetism. A phenomenon is not explained by being dwarfed. A fact is not transformed into a theory by being looked at through an inverted telescope. The hypothesis of ultimate indestructible atoms is not a necessary implication of the persistence of weight, and can at best account for the indestructibility of matter if it can be shown that there is an absolute limit to the compressibility of matter—in other words, that there is an absolutely least volume for every determinate mass. This brings us to the consideration of that general property of matter which probably, in the minds of most men, most urgently requires the assumption of atoms—its impenetrability.

"Two bodies can not occupy the same space"—this is the ordinary statement of the fact in question. Like the indestructibility of matter, it is claimed to be a datum of experience. "That all bodies are impenetrable," says Sir Isaac Newton, "we gather, not from reason, but from sense."* Let us see in what sense and to what extent this claim is legitimate.

The proposition, according to which a space occupied by one body can not be occupied by another, implies the assumption that space is an absolute, self-measuring, objective entity, and the further assumption that there is a least space which a given body will absolutely fill so as to exclude any other body. A verification of this proposition by experience, therefore, must amount to proof that there is an absolute limit to the compressibility of all matter whatsoever. Now, does experience authorize us to assign such a limit? Assuredly not. It is true that in the case of solids and liquids there are practical limits beyond which compression by the mechanical means at our command is impossible; but even here we are met by the fact that the volumes of fluids, which effectually resist all efforts at

* "*Corpora omnia impenetrabilia esse, non ratione, sed sensu colligimus.*"—Phil. Nat. Princ. Math., lib. iii, reg. 3.

further reduction by external pressure, are readily reduced by mere mixture. Thus, sulphuric acid and water at ordinary temperatures do not sensibly yield to pressure; but, when they are mixed, the resulting volume is materially less than the aggregate volumes of the liquids mixed. But, waiving this, as well as the phenomena which emerge in the processes of solution and chemical action, it must be said that experience does not in any manner vouch for the impenetrability of matter in all its stages of aggregation. When gases are subjected to pressure, the result is simply an increase of the expansive force in proportion to the pressure exerted, according to the law of Boyle or Mariotte (the modifications of and apparent exceptions to which, as exhibited in the experimental results obtained by Regnault and others, need not here be stated, because they do not affect the argument). A definite experimental limit is reached in the case of those gases only in which the pressure produces liquefaction or solidification. The most significant phenomenon, however, which experience contributes to the testimony on this subject, is the diffusion of gases. Whenever two or more gases which do not act upon each other chemically are introduced into a given space, each gas diffuses itself in this space as though it were alone present there; or, as Dalton, the reputed father of the modern atomic theory, expresses it, "Gases are mutually passive, and pass into each other as into vacua."

Whatever reality may correspond to the notion of the impenetrability of matter, this impenetrability is not, in the sense of the atomists, a datum of experience.

Upon the whole, it would seem that the validity of the first proposition of the atomic theory is not sustained by the facts. Even if the assumed unchangeability of the supposed ultimate constituent particles of matter presented itself, upon its own showing, as more than a bare reproduction of an observed fact in the form of an hypothesis, and could be dignified with the name of a generalization or of a theory, it would still be obnoxious to the criticism that it is a generalization from facts crudely observed and imperfectly apprehended.

In this connection it may be noted that the atomic theory has become next to valueless as an explanation of the impenetrability of matter, since it has been pressed into the service of the undulatory theories of radiance, and assumed the form in which it is now held by the majority of physicists, as we shall presently see. According to this form of the theory, the atoms are either mere points, wholly without extension, or their dimensions are infinitely small as compared with the distances between them, whatever be the state of aggregation of the substances into which they enter. In this view the resistance which a body, i. e., a system of atoms, offers to the intrusion of another body is due, not to the rigidity or unchangeability of volume of the individual atoms, but to the relation between the attractive and repulsive forces with which they are supposed to be endowed. There are physicists holding this view who are of opinion that the atomic constitution of matter is consistent with its penetrability—among them M. Cauchy, who, after defining atoms as "material points without extension," uses this language: "Thus, this property of matter which we call impenetrability is explained, when we consider the atoms as material points exerting on each other attractions and repulsions which vary with the distances that separate them. . . . From this it follows that, if it pleased the author of Nature simply to modify the laws according to which the atoms attract or repel each other, we might instantly see the hardest bodies penetrating each other, the smallest particles of matter occupying immense spaces, or the largest masses reducing themselves to the smallest volumes, the entire universe concentrating itself, as it were, in a single point." *

2. The second fundamental proposition of the modern atomic theory avouches the essential discontinuity of matter. The advocates of the theory affirm that there is a series of physical phenomena which are inexplicable, unless we assume that the constituent particles of matter are separated by void interspaces. The most notable among these phenomena are the dispersion and polarization of light. The grounds upon which the assumption of a discrete molecular structure of matter is deemed indispensable for the explanation of these phenomena may be stated in a few words.

According to the undulatory theory, the dispersion of light, or its separation into spectral colors, by means of refraction, is a consequence of the unequal retardation experienced by the different waves, which produce the different colors, in their transmission through the refracting medium. This unequal retardation presup-

* "Ainsi, cette propriété de la matière que nous nommons impénétrabilité se trouve expliquée, quand on considère les atomes comme des points matériels qui exercent les uns sur les autres des attractions ou répulsions variables avec les distances qui les séparent. . . . Il résulte encore de ce qui précède, que s'il plaisait à l'auteur de la nature de modifier seulement les lois suivant lesquelles les atomes s'attirent ou se repoussent, nous pourrions voir, à l'instant même, les corps les plus durs se pénétrer les uns les autres, les plus petites parcelles de matière occuper des espaces démesurés, ou les masses les plus considérables se réduire aux plus petits volumes, et l'univers entier se concentrer pour ainsi dire en un seul point." Sept Leçons de Physique Générale, ed. Moigno, p. 38 *seq.*

poses differences in the velocities with which the various colored rays are transmitted through any medium whatever, and a dependence of these velocities upon the lengths of the waves. But, according to a well-established mechanical theorem, the velocities with which undulations are propagated through a continuous medium depend solely upon the elasticity of the medium as compared with its inertia, and are wholly independent of the length and form of the waves. The correctness of this theorem is attested by experience in the case of sound. Sounds of every pitch travel with the same velocity. If it were otherwise, music heard at a distance would evidently become chaotic; differences of velocity in the propagation of sound would entail a distortion of the rhythm, and, in many cases, a reversal of the order of succession. Now, differences of color are analogous to differences of pitch in sound, both reducing themselves to differences of wave-length. The lengths of the waves increase as we descend the scale of sounds from those of a higher to those of a lower pitch; and similarly, the length of a luminar undulation increases as we descend the spectral scale, from violet to red. It follows, then, that the rays of different color, like the sounds of different pitch, should be propagated with equal velocities, and be equally refracted; that, therefore, no dispersion of light should take place.

This theoretical impossibility of dispersion has always been recognized as one of the most formidable difficulties of the undulatory theory. In order to obviate it, Cauchy, at the suggestion of his friend Coriolis, entered upon a series of analytical investigations, in which he succeeded in showing that the velocities with which the several colored rays are propagated may vary according to the wave-lengths, if it be assumed that the æthereal medium of propagation, instead of being continuous, consists of particles separated by sensible distances.

By means of a similar assumption, Fresnel has sought to remove the difficulties presented by the phenomena of polarization. In ordinary light, the different undulations are supposed to take place in different directions, all transverse to the course or line of propagation, while in polarized light the vibrations, though still transverse to the ray, are parallelized, so as to occur in the same plane. Soon after this hypothesis had been expanded into an elaborate theory of polarization, Poisson observed that, at any considerable distance from the source of the light, all transverse vibrations in a continuous elastic medium must become longitudinal. As in the case of dispersion, this objection was met by the hypothesis of the existence of "finite intervals" between the æthereal particles.

These are the considerations, succinctly stated, which theoretical physics are supposed to bring to the support of the atomic theory. In reference to the cogency of the argument founded upon them, it is to be said, generally, that evidence of the discrete molecular arrangement of matter is by no means proof of the alternation of unchangeable and indivisible atoms with absolute spatial voids. But it is to be feared that the argument in question is not only formally, but also materially, fallacious. It is very questionable whether the assumption of "finite intervals" between the particles of the luminiferous æther is competent to relieve the undulatory theory of light from its embarrassments. This subject, in one of its aspects, has been thoroughly discussed by E. B. Hunt, in an article on the dispersion of light,* and the suggestions there made appear to me worthy of serious attention. They are briefly these:

M. Cauchy brings the phenomena of dispersion within the dominion of the undulatory theory, by deducing the differences in the velocities of the several chromatic rays from the differences in the corresponding wave-lengths by means of the hypothesis of definite intervals between the particles of the light-bearing medium. He takes it for granted, therefore, that these chromatic rays are propagated with different velocities. But is this the fact? Astronomy affords the means to answer this question.

We experience the sensation of white light when all the chromatic rays of which it is composed strike the eye simultaneously. The light proceeding from a luminous body will appear colorless, even if the component rays move with unequal velocities, provided all the colored rays, which together make up white light, concur in their action on the retina at a given moment; in ordinary cases it is immaterial whether these rays have left the luminous body successively or together. But it is otherwise when a luminous body becomes visible suddenly, as in the case of the satellites of Jupiter, or Saturn, after their eclipses. At certain periods, more than forty-nine minutes are requisite for the transmission of light from Jupiter to the earth. Now, at the moment when one of Jupiter's satellites, which has been eclipsed by that planet, emerges from the shadow, the red rays, if their velocity were the greatest, would evidently reach the eye first, the orange next, and so on through the chromatic scale, until finally the complement of colors would be filled by the arrival of the violet ray, whose velocity is supposed to be the least. The satellite, immediately after its emersion, would appear

* Silliman's Journal, 2d series, vol. vii, p. 364 *seq.*

red, and gradually, in proportion to the arrival of the other rays, pass into white. Conversely, at the beginning of the eclipse, the violet rays would continue to arrive after the red and other intervening rays, and the satellite, up to the moment of its total disappearance, would gradually shade into violet.

Unfortunately for Cauchy's hypothesis, the most careful observation of the eclipses in question has failed to reveal any such variations of color, either before immersion or after emersion, the transition between light and darkness taking place instantaneously and without chromatic gradations.

Astronomy points to several other phenomena which are equally at war with the doctrine of unequal velocities in the movements of the chromatic undulations. Fixed stars beyond the parallactic limit, whose light must travel more than three years before it reaches us, are subject to great periodical variations of splendor; and yet these variations are unaccompanied by variations of color. Again, the assumption of different velocities for the different chromatic rays is discountenanced by the theory of aberration. Aberration is due to the fact that, in all cases where the orbit of the planet, on which the observer is stationed, forms an angle with the direction of the luminar ray, a composition takes place between the motion of the light and the motion of the planet, so that the direction in which the light meets the eye is a resultant of the two component directions—the direction of the ray and that of the observer's motion. If the several rays of color moved with different velocities there would evidently be several resultants, and each star would appear as a colored spectrum longitudinally parallel to the direction of the earth's motion.

The allegation of a dependence of the velocity of the undulatory movements, which correspond to, or produce, the different colors, upon the length of the waves, is thus at variance with observed fact. The hypothesis of "finite intervals" is unavailable as a supplement to the undulatory theory; other methods will have to be resorted to in order to free this theory from its difficulties.*

The negative evidence here adduced against the supposition of an atomic or molecular constitution of the light-bearing medium is reënforced by positive evidence derived from a branch of the atomic theory itself—the modern science of thermo-dynamics. Maxwell has remarked, with obvious truth, that such a medium (whose atoms or molecules are supposed to penetrate the intermolecular spaces of ordinary substances) would be nothing more nor less than a gas, though a gas of

* Since the publication of Cauchy's "Mémoire sur la dispersion de la lumière" (Prag, 1836), the dependence of the dispersive powers of different substances upon their states of aggregation and chemical composition has been the subject of extensive experimental research; and the most prominent physicists (Briot, Holtzmann, Redtenbacher, C. Neumann, Ketteler) now look for an explanation of the phenomena of dispersion to the action of ponderable matter, or to the interaction between it and the æther. Cf. Briot, "Essai sur la théorie mathématique de la lumière" (Paris, Mallet-Bachelier, 1864), p. 89 *seq.*; Redtenbacher, "Dynamidensystem," p. 130 *seq.*; Ketteler, "Ueber den Einfluss der ponderablen Molekuele auf die Dispersion des Lichts," etc. (Pogg. Ann., vol. cxl, pp. 2 *seq.* and 177 *seq.* An electro-magnetic theory of light, suggested by the proximate equality of the velocities with which light and electro-magnetic disturbances appear to be propagated through air and other media, and by the action of a magnet (observed by Faraday) in turning the plane of polarization round the direction of the luminar ray as an axis, was broached by Clerk Maxwell in 1865, and has recently been set forth at some length in his "Treatise on Electricity and Magnetism," vol. ii, pp. 383 *seq.* This theory is now being developed by Helmholtz, Lorentz, Fitzgerald, J. J. Thomson, and Lord Rayleigh.

great tenuity, and that every so-called vacuum would in fact be full of this rare gas at the observed temperature and at the enormous pressure which the æther, in view of the functions assigned to it by the undulatory theories, must be assumed to exert. Such a gas, therefore, must have a correspondingly enormous specific heat equal to that of any other gas at the same temperature and pressure, so that the specific heat of every vacuum would be incomparably greater than that of the same space filled with any other known gas. This remarkable consequence is not only without experimental warrant, but—inasmuch as it would apply to all vacua, including the intermolecular spaces of ordinary bodies of whatever state of aggregation—is in effect a fatal aggravation of a peculiar difficulty of the molecular theory which is in itself formidable to the highest degree. In the third chapter * I have adverted to the fact that, when a body is heated, a part only of the energy communicated to it appears in the form of temperature, i. e. (in the sense of modern theories), of progressive motions of the molecules, the other part being expended in the production of vibratory or rotatory motions of their constituent elements. According to the kinetic theory of gases, this latter part, the internal energy, so called, increases with the number of variables or degrees of freedom in each molecule, and with it, therefore, the specific heat, i. e., the ratio of the whole energy to that of translation which produces expansion or pressure, and is thus exhibited as temperature. If the molecules were "material points" without internal mobility, or perfectly elastic and perfectly smooth spheres, the total energy would be available for the production of translatory motion, and no part of it would be con-

* *Supra*, p. 86.

verted into internal energy. But if the molecules, though perfectly elastic, are not perfect spheres—as they can not be, whenever they consist of several atoms each—the specific heat must at least be equal to a certain minimum assigned by the theory. Now, the specific heats of oxygen, nitrogen, and hydrogen (all which are diatomic, their molecules consisting of at least two atoms each), as experimentally ascertained from a comparison of their specific heats at constant pressure and at constant volume, fall short of this minimum. And this theoretical minimum would be very materially increased by the addition of the specific heat due to the intermolecular æther, if this were also of atomic or molecular constitution; the discrepancy between the theoretical postulates and the experimental data would be immeasurably widened.

3. The third proposition of the atomic hypothesis assigns to the atoms, which are said to compose the different chemical elements, determinate weights corresponding to their equivalents of combination, and is supposed to be necessary to account for the facts whose enumeration and discussion constitute the science of chemistry. The proper verification of these facts is of great difficulty, because they have generally been observed through the lenses of the atomic theory, and stated in its doctrinal terms. Thus the differentiation and integration of bodies are invariably described as decomposition and composition; the equivalents of combination are designated as atomic weights or volumes, and the greater part of chemical nomenclature is a systematic reproduction of the assumptions of atomism. Nearly all the facts to be verified are in need of preparatory enucleation from the envelopes of this theory.

The phenomena usually described as chemical composition and decomposition present themselves to observation thus: A number of heterogeneous bodies concur in definite proportions of weight or volume; they interact; they disappear, and give rise to a new body possessing properties which are neither the sum nor the mean of the properties of the bodies concurring and interacting (excepting the weight which is the aggregate of the weights of the interacting bodies); and this conversion of several bodies into one is accompanied, in most cases, by changes of volume, and in all cases by the evolution or involution of heat, or other forms of energy. Conversely, a single homogeneous body gives rise to heterogeneous bodies, between which and the body from which they originate the persistence of weight is the only relation of identity.

For the sake of convenience, these phenomena may be distributed into three classes, of which the first embraces the persistence of weight and the combination in definite proportions; the second, the changes of volume and the evolution or involution of energy; and the third, the emergence of a wholly new complement of chemical properties.

Obviously, the atomic hypothesis is in no sense an explanation of the phenomena of the second class. It is clearly and confessedly incompetent to account for changes of volume, temperature, or latent energy. And, with the phenomena of the third class, it is apparently incompatible. For, in the light of the atomic hypothesis, chemical compositions and decompositions are in their nature nothing more than aggregations and segregations of masses whose integrity remains inviolate. But the radical change of chemical properties, which is the result of all true chemical action, and serves to distinguish it from mere mechanical mixture or separation, evinces a thorough destruction of that integrity. It may be that the appearance of this incompatibility can be obliterated by the device of ancillary hypotheses; but that leads to an abandonment of the simplicity of the atomic hypothesis itself, and thus to a surrender of its claims to merit as a theory.

At best, then, the hypothesis of atoms of definite and different weights can be offered as an explanation of the phenomena of the first class. Does it explain them in the sense of generalizing them, of reducing many facts to one? Not at all; it accounts for them, as it professed to account for the indestructibility and impenetrability of matter, by simply iterating the observed fact in the form of an hypothesis. It is another case (to borrow a scholastic phrase) of illustrating *idem per idem*. It says: The large masses combine in definitely-proportionate weights because the small masses, the atoms of which they are multiples, are of definitely-proportionate weight. It pulverizes the fact, and claims thereby to have sublimated it into a theory.*

The truth is, as Sir William Thomson has observed, that "the assumption of atoms can explain no property of a body which has not previously been attributed to the atoms themselves."

The foregoing considerations do not, of course, detract from the merits of the atomic hypothesis as a graphic or expository device—as an aid to the repre-

* That the assumption of atoms of different specific gravities is, on the basis of the atomic theory itself, simply absurd, has already been shown (*supra*, p. 35). According to the mechanical conception, which underlies the whole atomic hypothesis, differences of weight are differences of density; and differences of density are differences of distance between the particles contained in a given space. But, in the atom there is no multiplicity of particles and no void space; hence differences of density or weight are impossible in the case of atoms.

sentative faculty in "realizing" the phases of chemical or physical transformation. It is a fact beyond dispute that chemistry owes a great part of its practical advance to its use, and that the structural formulæ founded upon it have enabled the chemist, not merely to trace the connection and mutual dependence of the various stages in the metamorphosis of "elements" and "compounds," so called, but in many cases (such as that of the hydrocarbon series in organic chemistry) successfully to anticipate the results of experimental research. The question, to what extent the atomic theory is still indispensable to the chemist as a "working hypothesis," is at this moment under vigorous discussion among men of the highest scientific authority, many of whom do not hesitate to indorse the declaration of Cournot (made many years ago) that "the belief in atoms is rather a hindrance than a help" * not only because, as Cournot complains, it interposes an impassable chasm between the phenomena of the inorganic and those of the organic world, but because even as a representation of the phases and results of the most ordinary chemical processes it is both inadequate and misleading. The modifications to which it has lately been found necessary to subject it, in order to meet the exigencies of the present state of chemical science—modifications exemplified in the doctrines of constant and varying atomicities or valences, of molecular or atomic enchainments, etc., with the attendant theories (propounded by Kékulé and others) of molecular impact—attest the difficulties encountered in the

* "En somme, pour l'harmonie générale du système de nos connaissances, par conséquent (autant que nous pouvons en juger) pour la plus juste perception de l'harmonie qui certainement existe dans l'ensemble des choses, la foi dans les atomes est plutôt un embarras qu'un secours." Cournot, Traité de l'Enchainement des Idées Fondamentales dans les Sciences et dans l'Histoire, i, 264 *seq.*

attempt to bring the atomic hypothesis into conformity with the theoretical requirements of the hour. And, in proportion as the attention of the modern chemist is directed to the transference and transformation of energy involved in every instance of chemical "composition" and "decomposition" no less than in every case of allotropic change, its ineptitude as a figurative adumbration of the real nature of chemical processes becomes more and more apparent.*

I propose next to consider one of the most notable applications of the atomic hypothesis to physics—the kinetic theory of gases.

* As an illustration of the disfavor with which the atomic hypothesis is coming to be regarded by distinguished chemists, I may be permitted to quote a passage from an essay by the late Sir Benjamin C. Brodie, Professor of Chemistry at Oxford: "I can not but say that I think the atomic doctrine has proved itself inadequate to deal with the complicated system of chemical fact which has been brought to light by the efforts of modern chemists. I do not think that the atomic theory has succeeded in constructing an adequate, a worthy, or even a useful representation of those facts." "On the Mode of Representation afforded by the Chemical Calculus as contrasted with the Atomic Theory." Chemical News, August, 1867, p. 72. It is but fair to add, however, that I am not in sympathy with Brodie's own theoretical scheme so far as I understand it.

CHAPTER VIII.

THE KINETIC THEORY OF GASES.—CONDITIONS OF THE VALIDITY OF SCIENTIFIC HYPOTHESES.

In the fourth chapter * I have already given an outline of the doctrine now generally known and accepted as the kinetic theory of gases. The assumptions of this theory are that a gaseous body consists of a great number of minute solid particles—molecules or atoms—in perpetual rectilinear motion, which, as a whole, is conserved by reason of the absolute elasticity of the moving particles, while the directions of the movements of the individual particles are incessantly changed by their mutual encounters or collisions. The colliding particles are supposed to act upon each other only within very small distances and for very short times before and after collision, their motion being free, and consequently rectilinear, in the intervals between such distances and times. The durations of the rectilinear motions in free paths are, moreover, assumed to be indefinitely large as compared with the durations of the encounters and of the mutual actions.

This theory was first advanced by Kroenig, † and has since been elaborated by Clausius, Maxwell, Boltz-

* *Supra*, p. 41.

† Pogg. Ann., vol. xcix, p. 315 *seq*. As is usual in such cases, prelusions of the theory have since been discovered in the writings of various older physicists—cf. P. Du Bois-Reymond in Pogg. Ann., vol. cvii, p. 490 *seq*.

mann, Stefan, Pfaundler, and other physicists of the highest note. As in the case of the atomic hypothesis generally, I propose for the present to discuss, not so much the logical warrant, as the scientific value, of the theory in question. To this end it will be necessary, however, first to ascertain the true nature and function of a scientific hypothesis—not only the criteria of its value, but also the conditions of its validity.

A scientific hypothesis may be defined in general terms as a provisional or tentative explanation of physical phenomena.* But what is an explanation in the true scientific sense? The answers to this question which are given by logicians and men of science, though differing in their phraseology, are essentially of the same import. Phenomena are explained by an exhibition of their partial or total identity with other phenomena. Science is knowledge; and all knowledge, in the language of Sir William Hamilton, † is a "unification of the multiple." "The basis of all scientific explanation," says Bain, ‡ "consists in assimilating a fact to some other fact or facts. It is identical with the generalizing process." And "generalization is only the apprehension of the One in the Many." # Similarly Jevons: ‖ "Science arises from the discovery of identity amid diversity," and ^ "every great advance in science consists in a great generalization pointing

* Wundt has lately (Logik, i, 403) sought to distinguish hypotheses from "anticipations of fact" and to restrict the term "hypothesis" to a sense which, notwithstanding its etymological warrant, is at variance with ordinary as well as scientific usage.

† Lectures on Metaphysics (Boston ed.), pp. 47, 48.

‡ Logic, ii (Inductive), chap. xii, § 2.

Hamilton, *l. c.*, p. 48.

‖ Principles of Science, i, p. 1.

^ *Ib.*, ii, p. 281.

out deep and subtle resemblances." The same thing is stated by the author just quoted in another place:* "Every act of explanation consists in detecting and pointing out a resemblance between facts, or in showing that a greater or less degree of identity exists between apparently diverse phenomena."

All this may be expressed in familiar language thus: When a new phenomenon presents itself to the man of science or to the ordinary observer, the question arises in the mind of either: What is it?—and this question simply means: Of what known, familiar fact is this apparently strange, hitherto unknown fact a new presentation—of what known, familiar fact or facts is it a disguise or complication? Or, inasmuch as the partial or total identity of several phenomena is the basis of classification (a class being a number of objects having one or more properties in common), it may also be said that all explanation, including explanation by hypothesis, is in its nature classification.

Such being the essential nature of a scientific explanation of which an hypothesis is a probatory form, it follows that no hypothesis can be valid which does not identify the whole or a part of the phenomenon, for the explanation of which it is advanced, with some other phenomenon or phenomena previously observed. This first and fundamental canon of all hypothetical reasoning in science is formally resolvable into two propositions, the first of which is that every valid hypothesis must be an identification of two terms—the fact to be explained and a fact by which it is explained; and the second that the latter fact must be known to experience.

Tested by the first of these propositions, all hypoth-

* Principles of Science, ii, p. 166.

eses are futile which merely substitute an assumption for a fact, and thus, in the language of the schoolmen, explain *obscurum per obscurius*, or (the assumption being simply the statement of the fact itself in another form—the "fact over again") illustrate *idem per idem*. And the futility of such hypotheses goes to the verge of mischievous puerility when they replace a single fact by a number of arbitrary assumptions, among which is the fact itself. Some of the uses made of the atomic hypothesis, both in physics and chemistry, which have been discussed in the last chapter, afford conspicuous examples of this class of bootless assumptions; and similar instances abound among the mathematical formulæ that are not infrequently paraded as physical theories. These formulæ are in many cases simply results of a series of transformations of an equation which embodies an hypothesis whose elements are neither more nor less than the elements of the phenomenon to be accounted for, the sole merit of the emerging formula being that it is not in conflict with the initial one.*

* I hope not to be misunderstood as disparaging the services for which physical science is indebted to mathematics. These services—especially those rendered by modern analysis—are incalculable. But there are mathematicians who imagine that they have compassed a solution of all the mysteries involved in a case of physical action when they have reduced it to the form of a differential expression preceded by a group of integral signs. Even when their equations are integrable they should bear in mind that the operations of mathematics are essentially deductive, and, while they may extend, can never deepen a physical theory. Granting that mathematics are much more than καθάρματα ψυχῆς, and that their office in the investigation of the causes of natural phenomena is far more important than the purely regulative functions of formal logic in science generally—conceding that the application of mathematics to physics has not only brought to light the significance of many experimental results, but has often been a trustworthy guide to successful re-

In order to comply with the first condition of its validity, an hypothesis must bring the fact to be explained into relation with some other fact or facts by identifying the whole or a part of the former with the whole or a part of the latter. In this sense it has been well said that a valid hypothesis reduces the number of the uncomprehended elements of a phenomenon by at least one.* In the same sense it is sometimes said that

search—nevertheless some of our prominent physicists and mathematicians might still read with profit the ninety-sixth aphorism in the first book of Bacon's *Novum Organon:* "Naturalis Philosophia adhuc sincera non invenitur, sed infecta et corrupta; in Aristotelis scholâ per logicam; in Platonis scholâ per theologiam naturalem; in secundâ scholâ Platonis, Procli et aliorum *per Mathematicam, quae philosophiam naturalem terminare, non generare aut procreare debet.*" As to the value of the class of formulæ referred to in the text it may not be inappropriate to cite the words of Cournot (De l'Enchainement, etc., i, p. 249): "Tant qu'un calcul ne fait que rendre ce que l'on à tiré de l'observation pour l'introduire dans les éléments du calcul a vrai dire il n'ajoute rien aux donnécs de l'observation." To the same effect are the admirable reflections of M. Poinsot (Théorie Nouvelle de la Rotation des Corps, éd. 1851, p. 79): "Ce qui a pu faire illusion à quelques esprits sur cette espèce de force qu'ils supposent aux formules de l'analyse, c'est qu'on en retire, avec assez de facilité, des vérités déja connues, et qu'on y a, pour ainsi dire, soi-même introduites, et il semble alors que l'analyse nous donne ce qu'elle ne fait que nous rendre dans un autre langage. Quand un théorème est connu, on n'a qu'à l'exprimer par des équations; si le théorème est vrai, chacune d'elle ne peut manquer d'être exacte, aussi bien que les transformées qu'on en peut déduire; et si l'on arrive ainsi à quelque formule évidente, ou bien établie d'ailleurs, on n'a qu'à prendre cette expression comme un point de départ, à revenir sur ses pas, et le calcul seul paraît avoir conduit comme de lui-même au théorème dont il s'agit. Mais c'est en cela que le lecteur est trompé."

* "Der Verstand hat das Beduerfniss jede Erscheinung zu erklaeren d. h. dieselbe als das Resultat bekannter Kraefte oder Erscheinungen begrifflich abzuleiten. . . . Es geht hieraus hervor, dass jede Hypothese nur bekannte Kraefte oder Erscheinungen zur Erklaerung annehmen darf, indem die Annahme einer bisher unbekannten Kraft nur die Qualitaet des zu erklaerenden Phaenomen's aendern, aber nicht die Zahl der unerklaerten Momente reduciren kann. Soll eine Hypothese nicht vollkommen un-

every true theory or hypothesis is in effect a simplification of the data of experience—an assertion which must be understood, however, with due regard to the second proposition to be discussed presently, i. e., with the proviso that the theory be not a mere *asylum ignorantiæ*, of the kind denoted by the schoolmen as a *principium expressivum*, such as the explanation of the phenomena of life by reference to a *vital principle*, or of certain chemical processes by *catalytic action*. True scientific explanations are generally complicated in form, not only because most phenomena, on proper analysis, prove to be complex, but because the simplest fact is not the effect of a single cause, but the product of a great and often indeterminate multiplicity of agencies —the outcome of the concurrence of numerous conditions. The Newtonian theory of planetary motion is much more intricate than that of Kepler, according to which every planet is conducted along its path by an *angelus rector*; and the account given by modern celestial mechanics of the precession of the equinoxes is far less simple than the announcement that among the great periods originally established by the Author of the universe was the Hipparchian cycle. The old brocard, *simplex veri judicium*, is to be taken with many grains of allowance before it can be trusted as a safe rule in determining the validity or value of scientific doctrines.

I now come to the second requirement of the validity of an hypothesis: that the explanatory phenomenon (i. e., that with which the phenomenon to be explained is identified) must be a datum of experience. This

nuetz und demgemaess die Verstandesarbeit, welche sie zur Befriedigung eines Beduerfnisses erzeugte, keine zwecklose sein, *so muss jede Hypothese die Zahl der unbegriffenen Momente einer Erscheinung mindestens um eine erniedrigen.*" Zoellner, Natur der Kometen, p. 189 *seq.*

proposition is in substance equivalent to that part of Newton's first *regula philosophandi*,* in which he insists that the cause assigned for the explanation of natural things must be a *vera causa*—a term which he does not expressly define in the Principia, but whose import may be gathered from the following passage of his Opticks †: "To tell us that every species of things is endowed with an occult specific quality by which it acts and produces manifest effects is to tell us nothing. But to derive two or three general principles of motion *from phenomena* and afterward to tell us how the properties and actions of all corporeal things follow from these manifest principles would be a very great step in philosophy, though the causes of those principles were not yet discovered."

The requirement in question has long been the subject of animated discussion by J. S. Mill, Whewell, and others; but it will be found, I think, that, after making due allowance for necessary implications, there is little real disagreement among thinkers. The recent statement of G. H. Lewes ‡ that "an explanation to be valid must be expressed in terms of phenomena already observed," and the counter-statement of Jevons # that "agreement with fact (i. e., the fact to be explained) is the one sole and sufficient test of a true hypothesis," are both far too broad, and are, indeed, modified by Lewes and Jevons themselves in the progress of the discussion; but the claim of Mr. Lewes is nevertheless true in the sense that no explanation is real unless it is an identification of experiential data. The confusion which, as in so many other cases of scientific controversy, is at the bottom of the seeming disagreement

* Phil. Nat. Princ. Math., lib. iii.

† Fourth edition, p. 377.

‡ Problems of Life and Mind, ii, 7.

\# Princ. of Science, ii, 138.

between the contending parties, arises from a disregard of the circumstance that the identification of two phenomena may be both partial and indirect—that it may be effected by showing that the phenomena have some known feature in common on condition that the existence, in one or both of the phenomena, of some other feature not yet directly observed, and perhaps incapable of direct observation, be assumed. The aptest illustration of this is the much-debated undulatory theory of light. This hypothesis identifies light with other forms of radiance, and even with sound, by showing that all these phenomena have the element of vibration or undulation (which is well known to experience) in common, on the assumption of an all-pervading material medium, of a kind wholly unknown to experience, as the bearer of the luminar undulations. In this case, as in all similar cases, the identity lies, not in the *fictitious* element, the æther, but in the *real* element, the *undulation*. It consists, not in the *agent*, but in *the law of its action*. And it is obvious that every hypothesis which establishes coincidences between phenomena in particulars that are purely fictitious is wholly vain, because it is in no sense an identification of phenomena. It is worse than vain: it is meaningless—a mere collection of words or symbols without comprehensive import. As Jevons expresses it:* "No hypothesis can be so much as framed in the mind, unless it be more or less conformable to experience. As the material of our ideas is undoubtedly derived from sensation so we can not figure to ourselves any existence or agent but as endowed with some of the properties of matter. All that the mind can do in the creation of new existences is to alter combinations, or by anal-

* Princ. of Science, ii, 141.

ogy to alter the intensity of sensuous properties." J. S. Mill is, therefore, clearly wrong when he says * that, "an hypothesis being a mere supposition, there can be no other limits to hypotheses than those of the human imagination," and that "we may, if we please, imagine, by way of accounting for an effect, some cause of a kind utterly unknown and acting according to a law altogether fictitious." The unsoundness of the latter part of this proposition is evidently felt by Mill himself, for he adds at the end of the next sentence that "there is *probably* no hypothesis in the history of science in which both the agent itself and the law of its operation were fictitious." There *certainly* is no such hypothesis —at least none which has in any way subserved the interests of science.

An hypothesis may involve not only one but several fictitious assumptions, provided they bring into relief, or point to the probability, or at least possibility, of an agreement between phenomena in a particular that is real and observable. This is especially legitimate when the agreement thus brought to light is not between two, but a greater number of phenomena, and still more so when the agreement is not merely in one but in several real particulars between diverse phenomena, so that, in the language of Whewell,† "the hypotheses which were assumed for one class of cases are found to explain another of a different nature—a consilience of induction." An instance of this is afforded by the hypothesis just referred to of the luminiferous æther, which was at first believed also to explain the retardation of comets. But, while the probability of the truth of an hypothesis is in direct ratio to the number

* Logic, 8th ed., p. 394.

† History of the Inductive Sciences (Am. ed.), ii, 186.

of phenomena thus brought into relation, it is in the inverse ratio of the number of such fictions, or, more accurately, its improbability increases geometrically while the series of independent fictions expands arithmetically.* This finds illustration again in the undulatory theory of light. The multitude of fictitious assumptions embodied in this hypothesis, in conjunction with the failure of the consiliences by which it appeared at first to be distinguished, can hardly be looked upon otherwise than as a standing impeachment of its validity in its present form. However ready we may be to accede to the demands of the theorist when he asks us to grant that all space is pervaded, and all sensible matter is penetrated, by an adamantine solid exerting at each point in space an elastic force 1,148,000,000,000 times that of air at the earth's surface, and a pressure upon

* "En général," says Cournot (De l'Enchainement, etc., i, 103), "une théorie scientifique quelconque, imaginée pour relier un certain nombre de faits donnés par l'observation, peut être assimilée à la courbe que l'on trace d'après une loi géometrique, en s'imposant la condition de la faire passer par un certain nombre de points donnés d'avance. Le jugement que la raison porte sur la valeur intrinsèque de cette théorie est un jugement probable, une induction dont la probabilité tient d'une part à la simplicité de la formule théorique, d'autre part au nombre des faits ou des groupes des faits qu'elle relie, le même groupe devant comprendre tous les faits qui s'expliquent déjà les uns par les autres, indépendamment de l'hypothèse théorique. *S'il faut compliquer la formule à mesure que de nouveaux faits se révèlent à l'observation elle devient de moins en moins probable en tant que loi de la Nature ;* ce n'est bientôt plus qu'un échafaudage artificiel qui croule enfin lorsque, par un surcroit de complication, elle perd même l'utilité d'un système artificiel, celle d'aider le travail de la pensée et de diriger les recherches. Si au contraire les faits acquis à l'observation postérieurement à la construction de l'hypothèse sont reliés par elle aussi bien que les faits qui ont servi à la construire, si surtout des faits prévus comme conséquences de l'hypothèse reçoivent des observations postérieures une confirmation éclatante, la probabilité de l'hypothèse peut aller jusqu'à ne laisser aucune place au doute dans un esprit éclairé."

the square inch of 17,000,000,000,000 pounds *—a solid which, at the same time, wholly eludes our senses, is utterly impalpable and offers no appreciable resistance to the motions of ordinary bodies—we are appalled when we are told that the alleged existence of this adamantine medium, the æther, does not, after all, explain the observed irregularities in the periods of comets; that, furthermore, not only is the supposed luminiferous æther unavailable as a medium for the origination and propagation of dielectric phenomena, so that for these a distinct all-pervading electriferous æther must be assumed,† but that it is very questionable whether the assumption of a single æthereal medium is competent to account for all the known facts in optics (as, for instance, the non-interference of two rays originally polarized in different planes when they have been brought to the same plane of polarization, and certain phenomena of double refraction, in view of which it is necessary to suppose that the rigidity of the medium varies with the direction of the strain—a supposition discountenanced by the facts relating to the intensities of reflected light), and that for the adequate explanation of the phenomena of light it is "necessary to consider what we term the æther as consisting of two media, each possessed of equal and enormous self-repulsion or elasticity, and both existing in equal quantities throughout space, whose vibrations take place in perpendicular planes, the two media being mutually indifferent, neither attracting nor repelling." ‡ In this endless superfeta-

* Cf. Herschel, Familiar Lectures, etc., p. 282; F. De Wrede (President Royal Academy of Sciences in Stockholm), address, Phil. Mag., 4th ser., vol. xliv, p. 82.

† W. A. Norton, on Molecular Physics, Phil. Mag., 4th ser., vol. xxiii, p. 193.

‡ Hudson, on Wave Theories of Light, Heat, and Electricity, Phil.

tion of æthereal media upon space and ordinary matter, there are ominous suggestions of the three kinds of æthereal substances postulated by Leibnitz and Cartesius alike as a basis for their vortical systems. There is an impulsive whirl in our thoughts, at least, when we are called upon, in the interests of the received form of the undulatory theory, not only to reject all the presumptions arising from our common observation and all the analogies of experience, but to cumulate hypotheses and æthers indefinitely. And we are but partially reassured by the circumstance that the theory in question, besides accounting for the phenomena of optics which had been observed at the time of its promulgation, has the great merit of successful prevision, having led to the prediction of a number of facts subsequently discovered. These predictions, certainly, have not only been numerous, but several of them, such as Hamilton's announcement of conical refraction (afterward verified by Lloyd) and Fresnel's forecast (from

Mag. (iv), vol. xliv, p. 210 *seq.* In this article the author also points out the crudeness of the subsidiary hypotheses which have been framed to obviate other difficulties of the undulatory theory, among which are those discussed in the last chapter. "Waves of sound," he says, "in our atmosphere are 10,000 times as long as the waves of light, and their velocity of propagation about 850,000 times less, and, even when air has been raised to a temperature at which waves of red light are propagated from matter, the velocity of sound-waves is only increased to about double what it was at zero centigrade. Even their velocity through glass is 55,000 times less than the speed of the æthereal undulations, and the extreme slowness of change of temperature in the conduction of heat (as contrasted with the rapidity with which the vibrations of the æther exhaust themselves, becoming insensible almost instantly when the action of the existing cause ceases) marks distinctly the essential difference between molecular and æthereal vibrations. It appears to me, therefore, a very crude hypothesis to imagine a combination of æthereo-molecular vibrations as accounting for the very minute difference in the retardation of doubly refracted rays in crystals."

the imaginary form of an algebraic formula) of circular polarization after two internal reflections in a rhomb, are very striking. But, although anticipations of this sort justly serve to accredit an hypothesis, they are, as Mill has shown,* by no means absolute tests of their truth. Using the word "cause" in the sense in which it is commonly understood, an effect may be due to any one of several causes, and may, therefore, in many cases be accounted for by any one of several conflicting hypotheses, as becomes evident to the most cursory glance at the history of science. When an hypothesis successfully explains a number of phenomena with reference to which it was constructed, it is not strange that it should also explain others connected with them that are subsequently discovered. There are few discarded physical theories that could not boast the prevision of phenomena to which they pointed and which were afterward observed; among them are the one-fluid theory of electricity and the corpuscular theory of light.

There are, of course, other conditions of the validity of an hypothesis to which I have not yet adverted. Among them are those specified by Sir W. Hamilton, Mill, Bain, and others, such as that the hypothesis must not be contradictory of itself or in conflict with the known laws of nature (which latter requirement is, however, somewhat doubtful, inasmuch as the laws in question may be incomplete inductions from past experience to be supplemented by the very elements postulated by

* Logic, p. 356. Long before Mill, Leibnitz observed that success in explaining (or predicting) facts is no proof of the validity of an hypothesis, inasmuch as right conclusions may be drawn from wrong premisses—as Leibnitz expresses it, "comme le vrai peut être tiré du faux." Cf. Nouveaux Essais, chap. xvii, sec. 5—Leibnitii, opp. ed. Erdmann, p. 397.

the hypothesis); that it must be of a nature to admit of deductive inferences, etc. Upon all these it is not necessary, in view of my present purpose, to dilate. The two conditions which I have sought to enforce and illustrate are, in my judgment, sufficient tests of the validity and merits of the kinetic theory of gases.

The fundamental fact to be accounted for by this theory is that gases are bodies which, at constant temperatures and in the absence of external pressure, expand at even rate. From this fact the two great empirical laws, so called, expressive of those physical properties of a gas which are directly attested by experience, are the necessary and immediate consequences, being, indeed, nothing more than partial and complementary statements of it. The limitation of gaseous volume being produced by pressure alone—the cohibition of the bulk of a gas being due *solely* to pressure—it follows that it must be proportional to it; in other words, that the volume of a gas must be inversely as the pressure; and this is the law of Boyle or Mariotte. Again: temperature is measured by the uniform expansion of a column of gas (in the air-thermometer); hence, if all gases expand equally, temperature is proportional to the volume of a gas and conversely; this is the law of Charles.*

* One of the strangest incidents in the history of physics is the grave discussion of the question respecting the true law of gaseous expansion. "According to Gay-Lussac," says Balfour Stewart (Treatise on Heat, p. 60), "the augmentation of volume which a gas receives when the temperature increases 1° is a certain fixed proportion of *its initial volume at 0° C.*; while, according to Dalton, a gas at any temperature increases in volume for a rise of 1° by a constant fraction of *its volume at that temperature*. . . . The dilatation of gases has since been investigated by Rudberg, Dulong and Petit, Magnus and Regnault, and the result of their labors leaves little doubt that Gay-Lussac's method of expressing the law is much nearer the truth than Dalton's." Inasmuch as the experiments

The foregoing real definition (i. e., exhibition of the properties) of a gas applies only to ideal or perfect gases. In actual experience we meet with no gas which, in the absence of pressure, expands with absolute uniformity; and for that reason we do not know experimentally of any gas behaving in strict conformity to the laws of Boyle and Charles. Moreover, we are unable directly to observe a gas which is wholly free from pressure; the datum of experience is simply that gases expand (other things being equal) in proportion to the diminution of the pressure to which they are subjected. But in the case of many gases—those which

of Rudberg and others were necessarily made on the supposition that the coefficient of expansion was the same for all gases (the question relating, not to the expansion of some particular gas, but of gases generally), and, as the standard temperatures were those of the air-thermometer, it would have been surprising, indeed, if the result had been confirmatory of Dalton's view. A thermometer is graduated by dividing a given length of a tube of even bore into equal parts. It is clear, therefore, that the increment of volume resulting from the expansion of the air in such a tube through one degree is a fixed part of a constant volume initially assumed, and not of a constantly increasing volume; and the same thing is, of course, true of any other gas if it expands at the same rate. Dalton's form of the law of expansion would yield the following remarkable series of equal ratios—in which the first represents the rate of expansion of air in the thermometer, and the others stand for the rate (or rather rates) of expansion of the gas under examination (a being the linear expansion of the air in the thermometer, v its initial volume, a' the corresponding expansion in the gas under examination, v' its initial volume): $\frac{a}{v} = \frac{a'}{v'} = \frac{a'}{v' + a'} = \frac{a'}{v' + 2a'} = \frac{a'}{v' + 3a'} = \frac{a'}{v' + 4a'}$, etc., etc. The attempts at an experimental solution of the question here referred to are suggestive, by the way, of a doubt as to the correctness of the prevailing systems of thermometry, which are founded on the assumption of equalities of volume-ratios in which one of the terms is constant while the other is variable, i. e., of fractions which have the same numerator, but different denominators. These suggestions are but imperfectly met by the reflection that the bores of our thermometrical tubes are very small.

are either wholly incoercible, or coercible (i. e., reducible to the liquid or solid state) with great difficulty, and of nearly all gases at very high temperatures—the deviation from uniformity of expansion is very slight.

Now, how does the kinetic theory of gases explain the experiential fact or facts just stated? It professes to explain them on the basis of at least three arbitrary assumptions, not one of which is a datum of experience, viz.:

1. That a gas is composed of solid particles which are indestructible and of constant mass and volume.

2. That these constituent particles are absolutely elastic.

3. That these particles are in perpetual motion, and, except at very small distances, in no wise act upon each other, so that their motions are absolutely free and therefore rectilinear.

I refrain from adding a fourth assumption—that of the absolute equality of the particles, in mass at least—because it is claimed (though unjustifiably) to be a corollary from the other assumptions.

The first of these assumptions has been sufficiently considered in the last chapter. The second assumption asserts the absolute elasticity of the constituent solid particles. What is the import and scope of this assumption? The elasticity of a solid body is that property by means of which it occupies, and tends to occupy, portions of space of determinate volume and figure, and therefore reacts against any force or stress producing, or tending to produce, an alteration of such volume or figure with a counter-force or stress which, in the case of perfect elasticity, is exactly proportional to the altering force. Now, it is seen at once that the property—the *fact*—thus assumed in the constituent solid includes the very fact to be accounted for in the gas. A perfect gas reacts against a stress tending to reduce its volume with a spring proportional to the stress; and for this reason gases are defined as elastic fluids. This resilience of the gas against diminution of volume is obviously a simpler fact than the rebound of a solid against both diminution *and increase* of volume, *in addition to the reaction against a change of figure.* The resistance to *several* kinds of change implies a greater number of forces, and is therefore a more complex phenomenon, than the resistance to *one* kind of change.*

It thus appears that the presupposition of absolute elasticity in the solids, whose aggregate is said to constitute a gas, is a flagrant violation of the first condition of the validity of an hypothesis—the condition which requires a reduction of the number of unrelated elements in the fact to be explained, and therefore forbids a mere reproduction of this fact in the form of an assumption, and *a fortiori* a substitution of several arbitrary assumptions for one fact. Manifestly the explanation offered by the kinetic hypothesis, in so far as its second assumption lands us in the very phenomenon from which it starts, the phenomenon of resilience, is (like the explanation of impenetrability, or of the com-

* It may be said that the greater simplicity of the properties of a gas is purely conceptual. The identification of concepts with facts is undoubtedly the great fundamental error of speculation; but we are now dealing with the conceptual elements of the hypothesis under discussion. The opinion that a solid of constant volume (or, more accurately expressed, of variable volume, expanding or contracting to a fixed volume *proprio motu*) is a simpler thing than a uniformly expanding body is certainly not based upon any fact of experience, but is a mere prejudice of the intellect akin to the notion that a body at rest is a simpler phenomenon than a body in uniform motion, and generally that rest is simpler than motion. This prejudice has its root in our habitual oblivion of the essential relativity of all phenomena, which will be discussed hereafter.

bination of elements in definite proportions by the atomic theory) simply the illustration of *idem per idem*, and the very reverse of a scientific procedure. It is a mere *versatio in loco*—movement without progress. It is utterly vain; or rather, inasmuch as it complicates the phenomenon which it professes to explicate, it is worse than vain—a complete inversion of the order of intelligence, a resolution of identity into difference, a dispersion of the One into the Many, an unraveling of the Simple into the Complex, an interpretation of the Known in terms of the Unknown, an elucidation of the Evident by the Mysterious, a reduction of an ostensible and real fact to a baseless and shadowy phantom.*

Waiving the question already discussed, whether or not the assumed absolute solidity and constancy of volume of the supposed constituent particles are consistent (in the light of the mechanical theory generally) with their absolute elasticity, I proceed to consider the third assumption of the kinetic hypothesis. This assumption is an unavoidable supplement to the initial theoretical

* All theorists who attempt to account for a physical fact by a multiplication of arbitrary assumptions in which the fact itself is reproduced are liable to Aristotle's acute animadversion upon the Platonic doctrine of ideas—their endeavors are as nugatory as those of a person who, for the purpose of facilitating the operation of counting, begins by multiplying his numbers—*οἱ δε τας ἰδέας αἰτίας τιθέμενοι πρῶτον μεν ζητοῦντες τωνδὶ τῶν ὄντων λαβεῖν τᾶς αἰτίας ἕτερα τούτοις ἴσα τον αριθμὸν ἐκόμισαν ὥσπερ εἴ τις ἀριθμῆσαι βουλόμενος ἐλαττόνων μὲν ὄντων οἴοιτο μὴ δυνήσασθαι πλείω δε ποιήσας ἀριθμοίη.* Met., A. 9, 990, *et seq.* Occam's rule "*Entia non sunt multiplicanda praeter necessitatem*" has its applications in physics no less than in metaphysics; and there are physical theories of which Michel Montaigne, if he lived to-day, would say what he said of certain scholastic vagaries, three hundred years ago: "*On eschange un mot pour un aultre mot, et souvent plus incogneu. . . . Pour satisfaire à un doubte, ils m'en donnent trois; c'est la teste d'Hydra. . . . Nous communiquons une question; on nous en redonne une ruchée.*" Essais, iii, 13.

complication of the phenomenon of elasticity, produced by the arbitrary substitution of the resilience of a solid against increase or diminution of volume and change of figure for the reaction of a gas against diminution of volume alone. To get rid of one gratuitous feature of the hypothesis (the addition of the rebound against dilatation and distortion to that against compression) and to bring it into conformity with the fact to be explained, it becomes necessary to add another arbitrary feature—to endow the parts with incessant rectilinear motion in all directions. In respect to this assumption, which, like other assumptions of the mechanical theory, is based upon a total disregard of the relativity and consequent mutual dependence of natural phenomena, it is to be said, for the present, that it is utterly gratuitous, and not only wholly unwarranted by experience, but out of all analogy with it. Bodies which, except on the very verge of immediate contact, move independently without mutual attraction or repulsion or any sort of mutual action and thus present perfect realizations of the abstract concept of free and ceaseless rectilinear motion, are unheard-of strangers in the wide domain of sensible experience. So complete an abandonment of the analogies of experience is all the more surprising in view of the circumstance that the atomic hypothesis, whereof the kinetic theory of gases is a branch, is confessedly a concretion of suggestions derived from celestial mechanics. There is hardly a treatise on modern physics in which the atoms or molecules are not compared to planetary or stellar systems. "A compound atom," says Jevons,* "may perhaps be com-

* Principles of Science, i, 453. In Arwed Walter's Untersuchungen ueber Molecularmechanik, p. 216, the system of Jupiter and his satellites is called a "planetary molecule."

pared with a stellar system, each star a minor system in itself." But the bodies with which celestial mechanics deal are all subject to the law of attraction; and the import of the very first theorem of Newton's Principia is, that these bodies, if their motions are at any moment out of the same straight line, can never collide, but must always move in curved orbits at a distance from each other. Oblique impacts between them productive of rotations as well as of deviations from their paths before impact, as they are imagined by Clausius and the other promoters of the kinetic theory, are impossible. And this is true, not only when the mutual actions of the bodies vary inversely as the squares of their distances, but whenever they vary as any higher power of these distances—a proposition to be borne in mind in view of certain speculations of Boltzmann, Stefan, and Maxwell, of which I shall presently speak.

There is another very extraordinary and, in the light of all the teachings of science, unwarrantable feature in the assumption respecting the movements of the alleged solid constituent particles. I allude to the absolute discontinuity between the violent mutual action attributed to these particles during the few instants of time before and after their collisions, and their total freedom from mutual action during the comparatively long periods of their rectilinear motion along "free paths." And this leads me to say a few words in regard to certain subsidiary assumptions made by Maxwell and others in order to account for the anomalies exhibited by gases of different degrees of coercibility in their deviations from Boyle's and Charles's law. Maxwell assumes that the gas-molecules are neither strictly spherical nor absolutely elastic, and that their centers repel each other with a force inversely proportional to the fifth power of their distance;* while Stefan † endeavors to adjust the hypothesis to the phenomena in question by postulating that the molecules are absolutely elastic and perfect spheres whose diameters are inversely proportional to the fourth roots of the absolute temperatures of the gases. These assumptions, which are fatal to all claims of simplicity preferred on behalf of the kinetic hypothesis, are in no sense an outgrowth of its original postulates; both are purely gratuitous as well as without experiential analogy, and the first of them, that of Maxwell, is in direct defiance of all the inductions from the wide range of actual observation. They are both mere stop-gaps of the hypothesis, peace-offerings for its non-congruence with the facts, pure inventions to satisfy the emergencies created by the hypothesis itself.

It were work of supererogation to review in detail the logical and mathematical methods by which it is attempted, from an hypothesis resting on such foundations, to deduce formulæ corresponding to the facts of experience. I may be permitted to say, however, that the methods of deduction are only less extraordinary than the premisses. To account for the laws of Boyle and Charles resort is had to the calculus of probabilities, or, as Maxwell terms it,‡ the method of statistics. It is alleged that, although the individual molecules move with unequal velocities, either because these velocities were originally unequal, or because they

* Since this was written, Maxwell himself has abandoned this assumption as not conformable to the facts.

† Ueber die dynamische Diffusion der Gase. Sitzungsberichte der kaiserlichen Akademie der Wissenschaften, Mathem. naturw. Classe, vol. lxv, p. 323. Cf. also Boltzmann, Ueber das Wirkungsgesetz der Molekularkraefte, Sitzungsberichte, etc., vol. lxvi, p. 213.

‡ Theory of Heat, p. 288.

have become unequal in consequence of the encounters between them, nevertheless, there will be an average of all the velocities belonging to the molecules of a system (i. e., of a gaseous body) which Maxwell calls the "velocity of mean square." The pressure, on this supposition, is proportional to a product of the square of this average velocity into the number of molecules multiplied by the mass of each molecule. The product of the number of molecules into the mass of each molecule is then replaced by the density—in other words, the whole molecular assumption is, for the nonce, abandoned—and the velocity is eliminated as representing the temperature; it follows, of course, that the pressure is proportional to the density.

Similar procedures lead to the law of Charles and the "law" of Avogadro (according to which the number of molecules in any two equal volumes of gases of whatever kind is the same at the same temperatures and pressures—a law which is itself a mere hypothesis). It is claimed, on statistical grounds again, that not only the average velocity of a number of molecules in a given gaseous body is the same, but that "if two sets of molecules, whose mass is different, are in motion in the same vessel, they will, by their encounters, exchange energy with each other till the average kinetic energy of *a single molecule of either set* is the same." * "This," says Maxwell, "follows from the same investigation which determines the law of distribution of velocities in a single set of molecules." All this being granted, the law of Charles and the law of Avogadro (called by Maxwell the law of Gay-Lussac) are readily derived. And at the end of these devious courses of deduction Maxwell adds a disquisition on the properties

* Maxwell, *l. c.*, p. 289 *seq.*

of molecules, in which he claims to have made it evident that the molecules of the same substance are "unalterable by the processes which go on in the present state of things, and every individual of the same species is of exactly the same magnitude as though they had all been cast in the same mold, like bullets, and not merely selected and grouped according to their size, like small shot," and that, therefore, as he expresses it in another place,* they are not the products of any sort of evolution, but, in the language of Sir John Herschel, "have the essential character of manufactured articles."

Now, on what logical, mathematical, or other grounds is the statistical method applied to the velocities of the molecules in preference to their weights and volumes? What reason is given, or can be given, why the masses of the molecules should not be subjected to the process of averaging as well as their motions? None whatever. And, in the absence of such reason, the deductions of the kinetic theory, besides being founded on rickety premisses, are delusive paralogisms.

Upon these considerations I do not hesitate to declare that the kinetic hypothesis has none of the characteristics of a legitimate physical theory. Its premisses are as inadmissible as the reasoning upon them is inconclusive. It postulates what it professes to explain; it is a solution in terms more mysterious than the problem—a solution of an equation by imaginary roots of unknown quantities. It is a pretended explanation, of which it were unmerited praise to say that it leaves the facts where it found them, and is obnoxious to the old Horatian stricture: *nil agit exemplum, litem quod lite resolvit.*

* Bradford Lecture on the Theory of Molecules, cf. Popular Science Monthly, January, 1874.

Much is said about the support derived by the kinetic theory of gases from the revelations of the spectroscope. The spectra of gases, unlike those of solids and liquids, are not continuous, but consist of distinct colored lines or bands—showing, as is claimed, that in gases the vibrations of molecules do not interfere; that incandescent gases emit distinct kinds of light and not (as Jevons expresses it) luminar noises, because there is no clashing of the molecules disturbing the natural periods of vibration.* The spectroscope is, no doubt, the most important witness yet called on behalf of the kinetic theory; but the testimony of this witness is not all in its favor. "The spectroscope," says Maxwell himself,† "shows that some molecules can execute a great many different kinds of vibrations. They must, therefore, be systems of very considerable complexity, having far more than six variables. Now, every additional variable introduces an additional amount of capacity for internal motion without increasing the external pressure. Every additional variable, therefore, increases the specific heat, whether reckoned at constant pressure or constant volume. So does any capacity which the molecule may have for storing up energy in the potential form. But the calculated specific heat is already too great when we suppose the molecule to consist of two atoms only. Hence

* According to the latest interpretation of spectroscopic phenomena, the continuity or discontinuity of a spectrum is indicative, not so much of the state of aggregation, as of the molecular complexity of the body examined. It is said that a body yields a spectrum of lines when its molecules contain but a few atoms each; that, when they contain more, the spectrum presents the appearance of fluted bands; and that the spectrum is continuous when each molecule comprises a great number of atoms.

† On the Dynamical Evidence of the Molecular Constitution of Bodies, Nature, March 4 and 11, 1875, Nos. 279, 280.

every additional degree of complexity which we attribute to the molecule can only increase the difficulty of reconciling the observed with the calculated value of the specific heat."

It may seem strange that so many of the leaders of scientific research, who have been trained in the severe schools of exact thought and rigorous analysis, should have wasted their efforts upon a theory so manifestly repugnant to all scientific sobriety—an hypothesis in which the very thing to be explained is but a small part of its explanatory assumptions. But even the intellects of men of science are haunted by pre-scientific survivals, not the least of which is the inveterate fancy that the mystery by which a fact is surrounded may be got rid of by minimizing the fact and banishing it to the regions of the Extra-sensible. The delusion that the elasticity of a solid atom is in less need of explanation than that of a bulky gaseous body is closely related to the conceit that the chasm between the world of matter and that of mind may be narrowed, if not bridged, by a rarefaction of matter, or by its resolution into "forces." The scientific literature of the day teems with theories in the nature of attempts to convert facts into ideas by a process of dwindling or subtilization. All such attempts are nugatory; the intangible specter proves more troublesome in the end than the tangible presence. Faith in spooks (with due respect be it said for Maxwell's thermo-dynamical "demons" and for the population of the "Unseen Universe") is unwisdom in physics no less than in pneumatology.

CHAPTER IX.

THE RELATION OF THOUGHTS TO THINGS.—THE FORMATION OF CONCEPTS.—METAPHYSICAL THEORIES.

It has become evident, I take it, in the course of the preceding discussions, that, while modern physical science is professedly an endeavor to reduce the phenomena of nature to the elements of mass and motion, and thus to exhibit them as results or phases of mechanical action—claiming, on this ground, to be the only mode of dealing with these phenomena that is not in its nature metaphysical—nevertheless all the departments of science which have made decided advances beyond the first classificatory stage proceed upon assumptions and lead to consequences inconsistent with the object of this endeavor and with the fundamental principles of the mechanical theory. We find ourselves in the midst of a confusion, therefore, which is to be cleared up, if at all, by an inquiry into the origin of this theory and by a determination of its attitude toward the laws of thought and the forms and conditions of its evolution.

The account given, by ordinary psychologists and logicians, of the nature and operations of thought may, so far as it bears upon the matter now under consideration, be compressed into a few sentences. Thought, in its most comprehensive sense, is the establishment or recognition of relations between phenomena. Foremost among these relations—the foundation, in fact, of all others, such as those of exclusion and inclusion, coexistence and sequence, cause and effect, means and end—are the relations of identity and difference. The difference between phenomena is a primary datum of sensation. The very act of sensation is based upon it. It is one of the many acute observations of Hobbes that "it is all one to be always sensible of the same thing and not to be sensible of anything."* "We only know anything," says J. S. Mill,† "by knowing it as distinguished from something else; all consciousness is of difference; two objects are the smallest number required to constitute consciousness; a thing is only seen to be what it is by contrast with what it is not."

While the apprehension of phenomenal difference (which, however, may be, and in most cases is, replaced by its reproduction in memory) is the basis or prerequisite of thought, thought proper, i. e., discursive thought, begins with the apprehension of identity amid phenomenal difference. Objects are *perceived* as different; they are *conceived* as identical by an attention of the mind to their point or points of agreement. They are thus classified, the points of agreement, i. e., those properties of the objects of cognition which belong to them in common, serving as the basis of classification. When the number of objects classified is great, and some of these objects have more properties in common than others, a series of classes is formed. The objects are first divided into groups (called by the logicians *infimæ species*) severally embracing such objects as are characterized by the greatest number of common prop-

* "Sentire semper idem et non sentire ad idem recidunt." Hobbes, Physica, iv, 25 (opp., ed. Molesworth, vol. i, p. 321.

† Examination of Sir William Hamilton's Phil. (Am. ed.), vol. i, p. 14.

erties consistent with their difference; these groups are then collected and distributed into higher groups or species having a less number of properties in common, and so on, until we arrive at the least number of properties in which all the objects embraced in (*logicè* subsumed under) the *infimæ species* and the intermediate species agree, so as to characterize the highest class, or *summum genus.*

From this it follows that, in proportion as we ascend the scale of classification from the *infimæ species* to the *summum genus*, the number of objects embraced in the successive classes (species or genera) increases, while the number of characteristic properties decreases. Now, the complement of properties characteristic of a particular class is termed a *concept;* the number of objects denoted by each concept is called its *extension* or *breadth;* and the number of properties (which, as constituents of a concept, bear the name of attributes) connoted by it its *intension*, *comprehension* or *depth;* whence springs the law of logic that, the greater the extension of a concept, i. e., the greater the number of objects denoted, the less its comprehension, i. e., the number of attributes connoted; or, expressed with mathematical accuracy, that the extension varies in geometrical ratio inversely as the comprehension varies in arithmetical ratio.*

It is readily seen that the ascent from a lower (more comprehensive, but less extensive) to a higher (more extensive but less comprehensive) class is effected by a progressive segregation and ideal union of those attributes which the respective classes have in common; and this process is termed abstraction.

* For an exact statement of the law in question, see Drobisch, Neue Darstellung der Logik, Logisch-mathematischer Anhang (third ed., p. 206).

In the sense of the foregoing exposition, thought proper has been defined as "the act of knowing or judging of things by means of concepts,"* a concept being "a collection of attributes united by a sign and representing a possible object of intuition."† This definition of a concept, however, is obnoxious to criticism, as being either too wide or too narrow. It may be said, on the one hand, to be too wide: for it applies to the total array of attributes constituting the mental representation of a single object, without reference to the question whether or not they are shared by any other object, as well as to the factitious selection or collection of attributes characteristic of a class, i. e., of a plurality of objects. In other words, it is a definition of *singular concepts* (expressed by singular terms) as well as of *general concepts* (expressed by general terms, or, as Mill would say, class names). In the language of the old logicians, it includes *infimæ species*, and may stand for any singular object or singular quality, irrespective of the fact or degree of its generality. This criticism would be avoided by defining a concept, with Sir William Hamilton,‡ as "the cognition of the general character, point or points in which a plurality of objects coincide." On the other hand, the word "concept" is very generally employed in a sense for which Mansel's definition is too narrow. German logicians, for example, habitually designate not only every mental reproduction of a presentation of sense, in so far as it is or may be an element of a judgment or logical proposition, as a concept (*Begriff*), but also the last result of any series of abstractions. And the last results of abstraction, the

* Mansel, Prolegomena Logica, p. 22.

† *Ib.*, p. 60.

‡ Lectures on Logic, p. 87.

summa genera, are excluded by the definition of Mansel. It is neither necessary nor practicable here to attempt a minute discussion of the questions arising upon these divergences in the use of terms; nor can I stop to weigh the objections recently urged by Tauschinsky, Lotze, Sigwart, Wundt and others to the theory of conception as founded upon classification or subsumption. The controversies on this head between the logicians of the old and those of the new school, as well as the interminable disputes between the nominalists and the conceptualists to which so large a space is devoted in the writings of J. S. Mill,* are in the main mere wars of words, and the points of disagreement are foreign to the investigation upon which I am about to enter. To one or two of these points I may have occasion to recur hereafter; for the present my brief summary of the incidents of logical conception is to serve only as a clew to the meaning of certain logical terms I am constrained to employ, whenever this meaning is not sufficiently apparent from the context.

Now, in any discussion of the operations of thought, it is of the utmost importance to bear in mind the following irrefragable truths, some of which—although all of them seem to be obvious—have not been clearly apprehended until very recent times:

1. Thought deals, not with things as they are, or are supposed to be, in themselves, but with our mental representations of them. Its elements are, not pure objects, but their intellectual counterparts. What is present in the mind in the act of thought is never a thing, but always a state or states of consciousness. However much, and in whatever sense, it may be con-

* Cf. Mill's Examination of Sir William Hamilton's Philosophy, chap. xvii.

tended that the intellect and its object are both real and distinct entities, it can not for a moment be denied that the object, of which the intellect has cognizance, is a synthesis of objective and subjective elements, and is thus primarily, in the very act of its apprehension and to the full extent of its cognizable existence, affected by the determinations of the cognizing faculty. Whenever, therefore, we speak of a thing, or a property of a thing, it must be understood that we mean a product of two factors neither of which is capable of being apprehended by itself. In this sense all knowledge is said to be relative.

2. Objects are known only through their relations to other objects. They have, and can have, no properties, and their concepts can include no attributes, save these relations, or rather, our mental representations of them. Indeed, an object can not be known or conceived otherwise than as a complex of such relations. In mathematical phrase: things and their properties are known only as functions of other things and properties. In this sense, also, relativity is a necessary predicate of all objects of cognition.

3. A particular operation of thought never involves the entire complement of the known or knowable properties of a given object, but only such of them as belong to a definite class of relations. In mechanics, for instance, a body is considered simply as a mass of determinate weight and volume (and in some cases figure), without reference to its other physical or chemical properties. In like manner each of the several other departments of knowledge effects a classification of objects upon its own peculiar principles, thereby giving rise to different series of concepts in which each concept represents that attribute or group of attributes—

that aspect of the object—which it is necessary, in view of the question in hand, to bring into view. Our thoughts of things are thus, in the language of Leibnitz, adopted by Sir William Hamilton, and after him by Herbert Spencer, *symbolical*, not (or, at least, not only) because a complete mental representation of the properties of an object is precluded by their number and the incapacity of the mind to hold them in simultaneous grasp, but because many (and in most cases the greater part) of them are irrelevant to the mental operation in progress.

Again: the attributes comprised in the concept of an object being the representations of its relations to other objects, and the number of these objects being unlimited, it follows that the number of attributes is also unlimited, and that, consequently, there is no concept of an object in which its cognizable properties are exhaustively exhibited. In this connection it is worthy of mention that the ordinary doctrinal statement of the relation of concepts to judgments is liable to serious objection. A judgment is said to be "a comparison of two notions (concepts), with a resulting declaration of their agreement or disagreement" (Whately), or "a recognition of the relation of congruence or confliction between two concepts" (Hamilton). Here it is assumed that the concepts preëxist to the act of judgment, and that this act simply determines the fact or degree of their congruence or confliction. But the truth is that every concept is the result of a judgment, or of a series of judgments, the initial judgment being the recognition of a relation between two data of experience. In most cases, indeed, a judgment is a collatión of two concepts; but every synthetic judgment (i. e., every judgment in which the predicate is more than a mere display of one or more of the attributes connoted by the subject) transforms both concepts which it brings into relation, by either amplifying or restricting their respective implications.* When a boy learns that "a whale is a mammal," his notions, both of a whale and of a mammal, undergo a material change. From the judgment of Thomas Graham that "hydrogen is a metal," both the term "hydrogen" and the term "metal" emerged with new meanings. The announcement by Sterry Hunt, that "just as solution is chemical combination so chemical combination is mutual solution," extended the concept "solution" as well as the concept "chemical combination."

It is apparent, from these considerations, that the concepts of a given object are terms or links in numberless series or chains of abstractions varying in kind and diverging in direction with the comparisons instituted between it and other objects; that the import and scope of any one of these concepts are dependent, not only on the number, but also on the nature of the relations with reference to which the classification of ob-

* That this did not escape the attention of Sir William Hamilton, notwithstanding his definition of a judgment, appears from the following passage of his Lectures on Logic (Am. ed., p. 84: "A concept is a judgment: for, on the one hand it is nothing but the result of a foregone judgment, or series of judgments, fixed and recorded in a word, a sign, and it is only amplified by the annexation of a new attribute through a continuance of the same process." Among German thinkers Herbart had a clear view of the same truth. "Die Ausbildung der Begriffe," he says (Lehrbuch zur Psychologie, § 189, Werke, vol. v, p. 130), "ist der langsame, allmaelige Erfolg des immer fort gehenden Urtheilens." In another place (*id. ib.*, § 78, Werke, v, 59): "Es fragt sich, ob die Begriffe im strengen logischen Sinn nicht vielmehr logische Ideale seien, denen sich unser logisches Denken mehr und mehr annaehern soll. . . . Es wird sich ueberdiess zeigen, dass die Urtheile es sind, wodurch die Begriffe dem Ideal mehr und mehr angenaebert werden, daher sie den letzten in gewissem Sinne vorangehen."

jects is effected; and that for this reason, too, all thoughts of things are fragmentary and symbolic representations of realities whose thorough comprehension in any single mental act, or series of acts, is impossible. And this is true, *a fortiori*, because the relations of which any object of cognition is the entirety, besides being endless in number, are also variable—because, in the language of Herakleitos, all things are in a perpetual flux.

All metaphysical or ontological speculation is based upon a disregard of some or all of the truths here set forth. Metaphysical thinking is an attempt to deduce the true nature of things from our concepts of them. Whatever diversity may exist between metaphysical systems, they are all founded upon the express or implied supposition that there is a fixed correspondence between concepts and their filiations on the one hand and things and their modes of interdependence on the other. This fundamental error is, in great part, due to a delusory view of the function of language as an aid to the formation and fixation of concepts. Roughly stated, concepts are the meanings of words; and the circumstance that words primarily designate things, or at least objects of sensation and their sensible interactions, has given rise to certain fallacious assumptions which, unlike the ordinary infractions of the laws of logic, are in a sense natural outgrowths of the evolution of thought (not without analogy to the organic diseases incident to bodily life) and may be termed structural fallacies of the intellect. These assumptions are:

1. That every concept is the counterpart of a distinct objective reality, and that hence there are as many things, or natural classes of things, as there are concepts or notions.

2. That the more general or extensive concepts and the realities corresponding to them preëxist to the less general, more comprehensive concepts and their corresponding realities; and that the latter concepts and realities are derived from the former, either by a successive addition of attributes or properties, or by a process of evolution, the attributes or properties of the former being taken as implications of those of the latter.

3. That the order of the genesis of concepts is identical with the order of the genesis of things.

4. That things exist independently of and antecedently to their relations; that all relations are between absolute terms; and that, therefore, whatever reality belongs to the properties of things is distinct from that of the things themselves.

By the aid of these preliminaries I hope to be able to assign to the mechanical theory its true character and position in the history of the evolution of thought. Before I proceed to this, however, it may not be without interest, in connection with the preceding inquiry into the relation between concepts and their corresponding objects, to consider the question which has long been the subject of eager debate, whether and to what extent conceivability is a test of possible reality. It is contended by J. S. Mill and his followers, that our incapacity of conceiving a thing is no proof of its impossibility; while Whewell and Herbert Spencer maintain (though not strictly in the same sense and on the same grounds) that what is inconceivable can not be real or true.* A trustworthy judgment on the merits of this

* The precise form of Spencer's test of truth, which he terms the "Universal Postulate," is the "Inconceivability of the Opposite." Expressed in the strict language of logic, his thesis is, that every proposition whose contradictory is inconceivable must be true. But, inasmuch as every negation of a proposition is the affirmation of its contradictory,

controversy can only be formed after a careful determination of the conditions of conceivability as indicated by the nature of the process of conception which I have attempted to describe.

It has been shown that all true conception consists in the establishment of relations of partial or total identity between the fact to be conceived and other known facts of experience. The first condition of conceivability, therefore, is that the thing or phenomenon in question be susceptible of classification, i. e., of total or partial identification with objects or phenomena previously observed.

A second and very obvious condition of conceivability is the consistency of the elements of the concept to be formed with each other. It is clear that two attributes, one of which is the negation of the other, can not simultaneously belong to the same subject and thus be parts of the same concept.

These two are the only conditions which are directly deducible from the theory of conception, and may, therefore, with some propriety be termed theoretical conditions. But there is a third, practical condition: the consistency of the new concept with previously-formed concepts bearing upon the same subject-matter. As I have said, this is a practical condition—not so much a condition of conceivability as of ready conceivability. For the old concepts may be defective or erroneous; the very concept with which they conflict may supplement or supplant, rectify or destroy them.

Now, it is easily seen that fulfillment of the first condition can not be a test of reality. Facts or phenomena may present themselves to observation which

this is equivalent to the general statement that whatever is inconceivable can not be true.

are wholly unlike any fact or phenomenon theretofore observed, or whose likeness to the prior data of experience has not yet been detected. The history of science is full of startling discoveries; every period of active research brings to light phenomena which are not only unlooked-for, but without apparent analogy to other known facts. In view of this Liebig said: "The secret of all those who make discoveries is that they regard nothing as impossible." *

Thus far, then, I agree with Mr. Mill. But I can not follow him when he also rejects compliance with the second condition as a criterion of possibility, and refuses or neglects to distinguish between the case of inconceivability by reason of the apparent or real incongruity of a new fact or phenomenon with the data of past experience and the very different case of inconceivability on the ground of inconsistency between the several elements of a proposed concept. He instances the concept "a round square" as one which we are unable to form, and alleges that this inability is due solely to the inveteracy of our experience. "We can not conceive a round square," he says,† "not merely because no such object has ever presented itself in our experience, for that would not be enough. Neither, for anything we know, are the two ideas in themselves incompatible. To conceive a body all black and yet all white, would only be to conceive two different sensations as produced in us simultaneously by the same object—a conception familiar to our experience—and we should probably be as well able to conceive a round square as a hard square, or a heavy square, if it were

* Annalen der Pharmacie, x, 179.

† Examination of the Philosophy of Sir William Hamilton, i, 88, Am. ed.

not that in our uniform experience, at the instant when a thing begins to be round, it ceases to be square, so that the beginning of the one impression is inseparably associated with the departure or cessation of the other. Thus our inability to form a conception always arises from our being compelled to form another contradictory to it."

Our inability to conceive a round square due to the fact "that in our uniform experience at the instant when a thing begins to be round it ceases to be square," and to the inseparable association between incipient roundness and departing squareness! Whether any one has ever had such experience as is here spoken of, I do not know; but, if he has, I am confident that, even after being reënforced by a large inheritance of ancestral experience in the light of the modern theory of evolution, it will prove insufficient to account for the inseparable association which Mill brings into play. The simple truth is, that a round square is an absurdity, a contradiction in terms. A square is a figure bounded by four equal straight lines intersecting at right angles; a round figure is a figure bounded by a curve; and the oldest definition of a curve is that of "a line which is neither a straight line nor made up of straight lines."

Mill's claim is, in effect, if not in express words, a denial of the validity of the laws of non-contradiction and excluded middle, or (as he himself would prefer to say) an assertion that the fundamental laws of logic are, like all so-called laws of nature, mere experiential inductions, uniformity of experience being their only warrant. But, if these laws are not absolutely and universally binding as constitutive principles of thought and speech—if the same thing may, at the same time, be and not be, and if its affirmation and denial are not strict alternatives—we are fairly landed in the regions of utter nonsense, where all thinking is at an end and all language without meaning. The laws in question are principles constitutive of, because they are tacit conventions preliminary to, distinct thought and intelligible speech; and they are no more to be suspended in favor of Mill's theory of inseparable association than to be abrogated in furtherance of Hegel's dialectic process.

It ought to be said that there are expressions in the same chapter of Mill's book, from which I have just quoted, which show that the author was very ill at ease in the presence of his own theory. For instance, he says: * "These things are literally inconceivable to us, our minds and our experience being what they are. Whether they would be inconceivable if our minds were the same, but our experience different, is open to discussion. A distinction may be made which I think will be found pertinent to the question. That the same thing should at once be and not be—that identically the same statement should be both true and false—is not only inconceivable to us, *but we can not conceive that it could be made conceivable.*"

How strange that sentences like these should come from the pen of John Stuart Mill! First he denies that inconceivability is, in any sense or in any case, a test of truth or reality; but then he says it may be otherwise if the inconceivability itself is inconceivable! That is to say: a witness is utterly untrustworthy; but, when he makes a declaration respecting his own trustworthiness, he ought to be believed!

The whole theory of inseparable association, as here advanced and applied by Mill, is simply groundless, it being impossible, under his theory, to know what the

* *Loc. cit.*, p. 88.

experience of his numerous readers has been, except again by experience which he can not have had, since most of these readers were utterly unknown to him. And all attempts to argue questions with any one on such a basis are supremely foolish, Mill being bound, by his own doctrine, to accept the answer, "My experience has been otherwise," as conclusive. Mill's theory is thus subversive of itself, and every earnest sentence he has ever written is its practical refutation.

In reference to the case of inconceivability just discussed, and others analogous to it, it is to be observed that much of the perplexity and confusion which is characteristic of the disputes between Mill and his antagonists arises from the failure of the disputants to discriminate between purely formal concepts and the mental representations of physical realities. There is a very wide distinction between the relation of a concept to the object of thought in mathematics, for example, and the corresponding relation between a concept of a material object and that object itself. In mathematics, as in all the sciences which are conversant about single relations or groups of relations established (and, within the limits of the constitutive laws of the mind, *arbitrarily* established) by the mind itself, certain concepts are exhaustive in the sense that they imply, if they do not explicitly exhibit, all the properties belonging to the object of thought. Not only the constituents of such an object, but also the laws of their interdependence, being determined by the intellect, a single concept may be expanded into a series of others. Thus, a parabola is a line every point in which is equidistant from a fixed point and a given straight line: that is one of its concepts. And in this all the properties of the parabola—that it is a conic section formed by cutting a cone parallel to one of its sides, that the area of any one of its segments is equal to two thirds of its circumscribed rectangle, etc.—are implied, and from it they may be deduced. One of its attributes is an implication of all the others. Our concepts of material objects, on the contrary, as I have shown, are never exhaustive, for their complement of attributes is of necessity both incomplete and variable. To what strange vagaries this confusion has given rise in other departments of speculation we shall see in a future chapter.

I come now to the third condition of conceivability: the consistency of the concept to be formed with previous concepts *in pari materiâ*. By far the greatest number of the cases of alleged inconceivability are traceable to a breach of this condition—to the incompatibility of new facts or views with our intellectual prepossessions. Accordingly, most of the cases adduced by Mill in support of his theory are taken from this class. But he does not always apprehend their true character, and most of them are very imperfectly, if at all, accounted for by his theory. One of his instances is that of the denial, once all but universal, of the possibility of antipodes, on the ground of their inconceivability. According to Mill, this inconceivability has now vanished; we not only readily conceive them as possible, but know them to be real. This is true enough; but it finds its explanation, not in the law of inseparable association to which it is referred by Mill, but in the fact that our ancestors held an erroneous concept of the action of gravity. They supposed that the direction in which gravity acted was an absolute direction in space; they did not realize that it was a direction toward the earth's center of gravity; *downward* to them meant something very different from the sense we attach to that word. With this erroneous concept they could not reconcile the fact that the force of gravity held our antipodes in position as well as ourselves; nor can we. But we have a juster concept of gravity, and the mode and direction of its action; the spurious notion with which the notion of antipodes was inconsistent has been removed, and the inconceivability of antipodes is at an end.

Similar observations apply to another example brought forward by Mill: the inability to conceive *actio in distans*, to which extended reference has already been made in a preceding chapter. This inability results from the inconsistency of this concept with the prevailing notions respecting material presence. If we reverse the proposition that a body acts where it is, and say that a body is where it acts, the inconceivability disappears at once. One of the wisest utterances on this subject is the saying of Thomas Carlyle (quoted by Mill himself in another place): "You say that a body can not act where it is not? With all my heart; but, pray where is it?" Of course, a reconstitution of our familiar concepts of material presence, in the sense here indicated, would preclude the mechanical construction of matter from elements absolutely limited, hard, unchangeable and separated from each other by absolutely void spaces.

It is hardly necessary to add that, generally speaking, the inconceivability of a physical fact arising from its incongruity with preconceived notions is no proof of its impossibility or want of reality. Intellectual progress consists almost wholly in the rectification or subversion of old ideas not a few of which are held to be self-evident during long intellectual periods. The instances already cited from Mill are apt illustrations of this; and they may be cumulated without limit. Until the discovery of the composition of water, of the true theory of combustion, and of the relative affinities of potassium and hydrogen for oxygen, it was impossible to conceive a substance which would ignite on contact with water, it being one of the recognized attributes of water—in other words, a part of the concept water—that it antagonized fire. This previous concept was spurious, and, when it had been destroyed, the inconceivability of a substance like potassium disappeared. Similarly, we are now unable to conceive a warm-blooded animal without a respiratory system, because we conceive the idiothermic condition of an animal organism to depend mainly on the chemical changes taking place within it, chief among which is the oxidation of the blood, which requires some form of contact between the blood and the air, and therefore some form of respiration. If, however, future researches should destroy this latter concept—if it should be shown that the heat of a living body may be produced in sufficient quantity by mechanical agencies, such as friction—a non-respiring warm-blooded animal would at once become conceivable.

While thus a physical phenomenon, however little we may be able to conceive it without violence to our familiar ideas, may be real, it is otherwise in the domain of the formal sciences, such as logic and mathematics. There we find concepts founded upon fundamental postulates or axiomatic truths with which all new concepts, to be valid, must be consistent. The fact is that, in the sphere of the ideal relations of space and time, the third condition of conceivability is at bottom identical with the second, inasmuch as there all minor concepts are, by implication at least, constituents of some higher, more comprehensive concept whose validity requires their consistency with each other. All this is equally true of those purely formal concepts which constitute the theoretical basis of some of the physical sciences, such as the general propositions of kinematics or phoronomics; within the limits of their proper application they are authoritative tests of possibility. And even among the physical truths based upon induction there are many whose universality is so well established as to afford strong, if not conclusive presumption against the legitimacy of concepts and the reality of alleged phenomena which would invalidate them.

The foregoing discussion of the question of conceivability as a test of truth is by no means exhaustive. There are topics connected with it upon which it is not my province to enter. One of these topics is the specification of the conditions under which the inconsistency between the elements of a proposed concept becomes apparent. In many cases the inconsistency is latent and emerges only upon thorough exhibition of all the implications of these elements and their colligation—upon an explication which is familiarly known as *reductio ad absurdum.* The procedure, in such cases, is in effect a reduction of the propositions into which the concept may be resolved to their last degree of homogeneity, so that the conflict between them, if it exists, becomes explicit. The details of this subject, however, belong to treatises on logic.

CHAPTER X.

CHARACTER AND ORIGIN OF THE MECHANICAL THEORY. —ITS EXEMPLIFICATION OF THE FIRST AND SECOND RADICAL ERRORS OF METAPHYSICS.

It is the distinct claim of modern physicists that the mechanical theory rests on the sure foundation of sensible experience, and is thus contradistinguished from metaphysical speculation, which is said (and, in the sense indicated in the preceding chapter, truly said) to be based on mere figments of the intellect. We have now arrived at a stage in our discussion where the validity of this claim may be examined.

The mechanical theory postulates *mass* and *motion* as the absolutely real and indestructible elements of all forms of physical existence. Ordinarily these elements are designated as *matter* and *force;* but this designation is plainly inaccurate. The action of force upon a body, in the light of the mechanical theory, is simply the transference of motion from one body to another; force, in the sense in which the word is here employed, is nothing else than motion under the aspect of its actual or possible transference. And its necessary complement, or, rather, its essential correlate—that which would remain if a body were divested of everything that is not a form of force, or mode of motion—is not *matter*, but *mass.*

Now, it is clear that motion in itself is not, and can

not be, an object of sensible experience. We have experiential knowledge of moving bodies, but not of pure motion. And it is equally clear that mass—or, to use the ordinary term, *inert matter*, or *matter per se*—can not be an object of sensible experience. Things are objects of sensible experience only by virtue of their action and reaction. As Leibnitz said, "Whatever does not act does not exist"—*quod non agit, non existit.* Mass is nothing whereof the senses have direct cognizance; it is not presented to them either as volume, or as solidity, or as impenetrability. The only knowledge we have of mass is derived from the fact that different velocities, or accelerations, or changes of motion, are produced in different bodies (which may be of the same volume and of the same degrees of solidity and impenetrability) by the action of the same force or the transference of the same motion. Apart from the atomic theory, mass is but another name for *inertia;* and this is known, measured, and determined solely by the amount of force or motion which must act upon, or be communicated to, a given body in order to produce in it a determinate velocity, or, more accurately and generally, a determinate rate of acceleration or deflection. Without its relation to and union with force or motion, it has no existence, just as force or motion has no existence without its relation to and union with inertia. The reality of either presents itself to experience as well as to thought only by means of the other.

The truth is, that neither mass nor motion is substantially real, but both are concepts, or, rather, constituents of a concept—the concept *matter.* They are ultimate products of generalization—the intellectual vanishing-points of the lines of abstraction which proceed from the *infimæ species* of sensible experience.

Matter is the *summum genus* of the classification of bodies on the basis of their physical and chemical properties. It is not, therefore, a real thing, but the ideal complement of two attributes belonging to all bodies alike. The two attributes are inseparable, not only in fact, but also in thought. When, in ascending the scale of classification, we have progressively dismissed from our mental representations of the several physical objects all the attributes whereby they differ, we reach at last two attributes wherein they agree, and which can not be sundered without transcending the limits within which the conception of physical reality is possible. They are both indispensable components of the highest concept under which any form of physical existence can be subsumed.

From this the true character of the mechanical theory is at once apparent. That theory takes, not only the ideal concept *matter*, but its two inseparable constituent attributes, and assumes each of them to be a distinct and real entity. And this identification of concepts with real, sensible objects, this confusion of abstractions with things, is one of the old fundamental errors of metaphysical speculation. It is the first of the fallacious assumptions of metaphysics enumerated in the last chapter.* The mechanical theory, in common with all metaphysical theories, hypostasizes partial, ideal, and, it may be, purely conventional groups of attributes, or single attributes, and treats them as varieties of objective reality. Its basis, therefore, is essentially metaphysical. The mechanical theory is, in fact, a survival of mediæval realism. Its substantial elements are legitimate logical descendants of the *universalia ante rem* and *in re* of the scholastics, differing from

* *Supra*, p. 137.

them, at most, in this, that they are summits of abstraction reached by ascents along gradations of sensible properties ascertained by observation and experiment, and not by escalades of the misty heights of traditional predicables representing early, crude, and vague fancies of the human intellect.

The metaphysical character of the mechanical theory appears, however, not only in its adoption of the first of the fallacious assumptions of all metaphysics, according to which each concept is the counterpart of a real thing, but also in the second of these assumptions, which is, as I have said,* that the more general or extensive concepts, and the realities corresponding to them, preëxist to the less general and more comprehensive concepts and their corresponding realities, and that the latter concepts and realities are derived from the former either by a successive addition of attributes or properties, or by a process of evolution, the attributes or properties of the former being taken as implications of those of the latter.

In the leading metaphysical systems, the order of reality is completely inverted. The *summa genera* of abstraction—the highest concepts—are deemed the most, and the data of sensible experience the least real of all forms of existence. The ground of this fancy is that the former, which include the properties common to all things, are assumed to constitute their substance, i. e., the permanent, invariable substratum of the properties by which particular things are distinguished, these being regarded, by reason of their variability, as mere accidents. According to the older view of the relation of the accidents to the substance, or of the characteristic attributes of the lower to those of the higher concepts,

* *Supra*, p. 138 *seq.*

the inferior concepts or realities are formed by a successive addition of attributes or properties to the higher concepts or realities; the varieties of objective reality are held to be due to a synthesis of substance and accidents; and this view may, therefore, be called the *synthetic* view. In contrast to this stands the later, *analytical* view, presented in evolutionary or pantheistic systems in which the lower conceptual or real forms are supposed to be contained or implied in the higher forms and to be derived from them by processes of evolution or development. All this has its exact analogue in the mechanical theory. Forty years ago the creed of an ordinary physicist was something like this: Primordially there existed, through an act of creation or from all eternity, myriads of hard and unchangeable material particles. There also existed certain forces equally unchangeable, such as the forces of attraction and cohesion, heat, electric, magnetic, chemical forces, and so on. To the constant or variable, partial or concurrent action of these forces upon the material particles are due all the phenomena of physical reality. In this action the material particles are the passive and the forces the active element; but these elements, of course, preëxist to the action. Matter in itself is passive, dead; all motion or life is caused by force; and the only possible solution of the problems of physiology, no less than those of physics and chemistry, consists in the enumeration of the forces acting upon the material particles and in the exact quantitative determination of the effects produced by their action.

In the main this creed is evidently a reproduction of the old synthetic view of metaphysics. And it is gradually giving way to a new doctrine which is similarly a reproduction of the metaphysical sequel which

I have termed the analytical or evolutionary view. The recent theories of the correlation and mutual convertibility of forces, as part of the principle of the conservation of energy, have shaken, if not destroyed, the notion of a multiplicity of independent original forces; and, moreover, physiologists like Du Bois-Reymond recognize force as the invariable concomitant, if not as the essential attribute or primary quality, of matter, asserting that to every constant primordial mass belongs a constant primordial quantity of force, and that all the transformations of matter are produced by a differentiation of this primordial force. From this the suggestion is natural that all the varieties of physical existence were potentially contained in and have been gradually evolved from matter in general, or matter *per se*.

In August, 1874, Professor Tyndall, then President of the British Association, delivered an inaugural address at a meeting of the Association at Belfast, in which he made the following declaration:

"Abandoning all disguise, the confession that I feel bound to make before you is that I prolong the vision backward across the boundary of the experimental evidence, and discern, in that matter which we, in our ignorance and notwithstanding our professed reverence for its Creator, have hitherto covered with opprobrium, the promise and potency of every form and quality of life."

This announcement gave rise to a commotion which was hardly justified by its tenor. For the solemnity of the avowal was somewhat out of proportion to its novelty. Tyndall's words were little more than a new wording of an old thought of Francis Bacon, who said, more than two centuries ago:

"And matter, whatever it is, must be held to be so adorned, furnished, and formed, that all virtue, essence, action, and motion may be the natural consequence and emanation thereof." *

And the same thing had been repeated, many times since, by the metaphysical evolutionists, in terms substantially like those of Schelling: "Matter is the general seed-corn of the universe wherein everything is involved that is brought forth in subsequent evolution." †

Nevertheless, Tyndall's statement is memorable and significant as indicating the changes which the mechanical theory is undergoing in the minds of modern physicists.

Tyndall is one of the most strenuous advocates of the atomo-mechanical theory and a persistent stickler for its dominant features. When he speaks of matter, he means a definite group of distinct and real atoms or molecules. "Many chemists of the present day," he said in another address (also delivered before the British Association, at Liverpool, and republished by him shortly before the Belfast meeting ‡) "refuse to speak of atoms and molecules as real things. Their caution leads them to stop short of the clear, sharp, mechanically-intelligible atomic theory enunciated by Dalton, or any form of that theory, and to make the doctrine of multiple proportions their intellectual bourn. I respect the caution, though I think it is here misplaced. The chemists who recoil from these notions of atoms and

* "*Atque asserenda materia (qualiscunque ea sit) ita ornata et apparata et formata, ut omnis virtus, essentia, actus atque motus naturalis ejus consecutio et emanatio esse possit.*" Baco, De Princ. atque Origg., Opp. ed. Bohn, vol. ii, p. 691.

† "*Die Materie ist das allgemeine Samenkorn des Universums, worin Alles verhuellt ist, was in spaeteren Entwickelungen sich entfaltet.*" Schelling, Ideen zu einer Philos. der Natur, 2d ed., p. 315.

‡ Fragments of Science (Am. ed.), p. 358.

molecules accept without hesitation the undulatory theory of light. Like you and me, they one and all believe in an æther and its light-producing waves. Let us consider what this belief involves. Bring your imagination once more into play and figure a series of sound-waves passing through air. Follow them up to their origin, and what do you there find? A definite, tangible, vibrating body. It may be the vocal chords of a human being, it may be an organ-pipe, or it may be a stretched string. Follow in the same manner a train of æther-waves to their source; remembering at the same time that your æther is matter, dense, elastic, and capable of motions subject to and determined by mechanical laws. What, then, do you expect to find as the source of a series of æther-waves? Ask your imagination if it will accept a vibrating multiple proportion —a numerical ratio in a state of oscillation.* I do not think it will. You can not crown the edifice by this abstraction. The scientific imagination, which is here authoritative, demands as the origin and cause of a series of æther-waves a particle of vibrating matter quite as definite, though it may be excessively minute, as that which gives origin to a musical sound. Such a particle we name an atom or a molecule. I think the seeking intellect, when focused so as to give definition without penumbral haze, is sure to realize this image at the last."

* When Tyndall wrote this he probably had before him W. K. Clifford's lecture delivered before the Royal Institution in 1867, in which occurred this passage: "In order to explain the phenomena of light, it is not necessary to assume anything more than a periodical oscillation between two states at any given point of space." (Clifford's Lectures and Essays, vol. i, p. 85.) Or the suggestion may have been taken from J. S. Mill, who, in a note to chapter xiv, book iii, of his Logic, referring to certain observations of Dr. Whewell, characterizes the imponderable æther as an "undulating agency."

The plain import of these sentences is, that an æthereal or other atom or molecule is related to its vibratory motion just as any ordinary body is related to its movements of translation—as a stellar or planetary body, for instance, is related to its movements of rotation or revolution; and that just as the conception of the stellar or planetary body of necessity precedes the conception of its rotatory or revolutionary motion, so also the conception of the atom or molecule of necessity precedes the conception of the vibratory motion whereof light, heat, electricity, chemical action, etc., are known, or supposed to be, modes. In other words: to make the existence of matter, such as we deal with in action and in thought, conceivable, we are constrained, according to Tyndall, to assume ultimate material particles as preëxisting to those motions or manifestations of force which are apprehended as light, heat, electricity, chemical action, etc. And what is true of the concept is true of the thing. The thing must *be*, before it can act or be acted upon, agreeably to the old maxim: *Operari sequitur esse.**

* It requires but little reflection to see that the realization of definite atoms or molecules, susceptible of, but preëxisting to motion, in the focus of Tyndall's "seeking intellect" is sheer delusion. Let us, for a moment, contemplate an ultimate particle of matter in its state of existence in advance of all its motion. It is without color, and neither light nor dark; for color and lightness are, according to the theory of which Tyndall is a distinguished champion, simply modes of motion. It is similarly without temperature—neither hot nor cold, since heat, also, is a mode of motion. For the same reason it is without electric, magnetic and chemical properties—in short, it is destitute of all those qualities in virtue of which, irrespective of its magnitude, it could be an appreciable object of sense, unless we except the properties of weight and extension. But weight is a mere play of attractive forces; and extension, too, is known to us only as resistance which, in turn, is a manifestation of force, a phase of motion. Thus the difficulty in grasping these primordial things lies, not in their excessive minuteness, but in their total destitution of quality. The solid, tangible reality craved by Tyndall's "scientific

This view, presented by Tyndall in his Liverpool address, is the old synthetic notion of metaphysical realism. The atoms or molecules are the substances existing in advance of their different modes of motion which are superinduced or added to them as their accidents. But in the Belfast address this view is (unconsciously, no doubt) so modified as to shade into the evolutionary or analytical aspect. Matter is now said to include or involve even the forms and qualities of life at the outset —to contain them, if not actually, at least potentially—so that they proceed from it by spontaneous development.

That all attempts to construct physical phenomena by a synthesis of hypostasized conceptual elements, under the first or synthetical view, are futile, in physics no less than in metaphysics, is now sufficiently evident upon considerations variously presented. Whether these elements be *substance* and *accident*, or *matter* and *force*, they are equally unreal, and no reality can be produced by their adjunction. And the fancied evolution of things, or lower, more comprehensive concepts from higher, more extensive concepts, in conformity with the second, analytical view, is also found to be delusive upon simple reference to the nature of the process of conception. Higher concepts are formed out of lower concepts by the omission or rejection of differential attributes; and there is nothing, certainly, in this logical process from which it can be legitimately inferred that the rejected attributes are contained or implied in those that are retained and in whose union the higher concepts consist.

imagination" is "*nec quid, nec quantum, nec quale*," and wholly vanishes from the "seeking intellect," the moment this intellect attempts to seize it apart from the motion which is said to presuppose it as its necessary substratum.

It is needless to say, I trust, that this nowise affects the validity of theories of evolution within the domains of real physical existence in their application to organic (and, within limits, to inorganic) forms. Questions of derivation and descent, and of organic and functional differentiation and distribution, are questions of fact to be determined in accordance with the data of observation and experiment. Modes of existence may be genetically connected, though there is no mutual implication of them, and though no form of physical reality is legitimately deducible from a concept. Aristotle's dictum, *ἐκ δε τῶν νοητῶν οὐδὲν γίνεται μέγεθος*, has a fuller meaning than that assigned to it by his scholastic disciples: things are not born of concepts. And, as will appear still more clearly in the next chapter, the filiation of concepts is not at all identical with the filiation of things.

The errors of evolutionism in its confessedly metaphysical forms (exhibited in numerous hylozoic and pantheistic doctrines) are more glaring, it is true, than those of materialistic evolutionism. It is characteristic of many of the most prominent metaphysical systems that the *summa genera* which serve as the basis of evolution are reached by leaps into vacuity beyond the boundaries of legitimate generalization. Thus Hegel evolves all things from pure *Being*, which, as he him self says, is wholly devoid of attributes—a mere logical phantom conjured up by a forced rejection of the last attributes that can be constitutive of the *summum genus* of any classification of phenomena whatever.* This

* Strictly speaking, the foundation of Hegel's "dialectic process" is not even a phantom of reality. "Being *per se*" is not so much as the mere locus of a vanished attribute. The copula between subject and predicate is nothing more than the formal expression of the fact that

phantom, as Hegel expressly declares, is not to be distinguished from, and therefore identical with, pure *Nothing;* and for this reason some of Hegel's intellectual descendants—Dellingshausen, Rohmer, Werder, George and others—have boldly undertaken to deduce the phenomenal world from the alleged concept *Nothing* or *Zero.* The same attempt is made by other metaphysicians in whose systems the initial blank appears under various disguises—by Schopenhauer and Hartmann, for instance, whose germinal principle is an impersonal will, a concept whose attributes are contradictory of each other, and which is, therefore, as void as the pseudo-concept Nothing. The most imposing among the disguises of the substantial Nothing as the fountain and origin of all phenomenal existence are *The Absolute* and *The Thing per se*, both of which are denials in terms of all possible relation, and thus negations of all possible attributes, inasmuch as every attribute is essentially a relation. But, although such concepts as *matter* and *force* are somewhat less hollow than the pseudo-concepts of current metaphysical speculations, they are not less unavailable as starting-points for the evolution of concrete physical realities.

Like all metaphysical theories, the mechanical theory, by its identification of concepts with things, has given rise to a number of false antagonisms and groundless discussions. One of the most noted controversies of the time is that between the champions of the *me-*

the relation of identity, inclusion or coexistence subsists between two attributes, or between an attribute and a group of attributes. It is a mere abstract line (or pair of lines) pointing from the generic to the differential constituents of a concept. "Pure Being" is simply the specter of the copula between an extinct subject and a departed predicate. It is a sign of predication which "lags superfluous on the stage" after both the predicate and that whereof it was predicated have disappeared.

chanical or *corpuscular* theory of matter, who assert that it is a real thing independent of force, and the defenders of the *dynamical* theory, who maintain that material particles are mere centers or spheres of force. The corpuscular doctrine is held by the majority of physicists in common with ordinary men, while the dynamical view—originally the outgrowth of metaphysical speculation—has been broached, on grounds that are alleged to be non-metaphysical, by Boscovich, Ampère, Faraday, and many others. Faraday's opinion is concisely stated by Tyndall: * "What do we know of the atom apart from its force? You imagine a nucleus which may be called *a*, and surround it by forces which may be called *m*; to my mind the *a* or nucleus vanishes and the substance consists of the powers *m*. And, indeed, what notion can we form of the nucleus independent of its powers? What thought remains on which to hang the imagination of an *a* independent of the acknowledged forces?"

When Faraday reasoned thus he was probably unaware that he but reproduced old reflections of Aristotle † which have since found frequent expression in the writings of modern thinkers,‡ of which the following may be taken as an example:

"It is a mere delusion of the phantasy that something, we know not what, remains after we have denuded an object of all the predicates belonging to it." #

* Faraday as a Discoverer, Am. ed., p. 123. For Faraday's own statement of this view, see his "Speculation touching Electric Conduction and the Nature of Matter," Phil. Mag., ser. iii, vol. xxiv, p. 136.

† De Gen. et Corrupt., ii, 1, 3, 4, 6; Met., iii, 5; iv, 2; vi, 1.

‡ Cf. i. a. Locke, Essay on Human Understanding, book ii, chapters xxiii and xxiv.

\# "*Es ist eine blosse Taeuschung der Einbildungskraft, dass, nachdem*

The antagonism thus presented is utterly baseless. Matter can no more be realized or conceived as mere passive, spatial presence, than as a concretion of forces. Force is nothing without mass, and mass is nothing without force. Just as the metaphysician can not conceive the "thing" or substance apart from its properties, or, conversely, the properties apart from the substance, so the physicist can not grasp matter (i. e., mass) without force, or force without matter. Mass, inertia, or matter *per se*, is indistinguishable from absolute nothingness; for mass reveals its presence or evinces its reality only by its action, its balanced or unbalanced force, its tension or motion. And, on the other hand, pure force is equally nothing; for if we reduce the mass upon which a given force, however small, acts, to its limit zero—or, mathematically expressed, until it becomes infinitely small—the consequence is that the velocity of the resulting motion is infinitely great, and that the "thing" (if under these circumstances we may still speak of a thing) is at any given moment neither here nor there, but everywhere—that there is no real presence. It is impossible, therefore, to construct matter by a synthesis of forces. And it is incorrect to say, with Bain,* that "matter, force, and inertia, are the three names for substantially the same fact," or, that "force and matter are not two things, but one thing," or,† that "force, inertia, momentum, matter, are all but one fact," the truth being that force and inertia are conceptual integrants of matter, and neither is in any proper sense a fact.

man einem Object die einzigen Praedikate die es hat, hinweggenommen hat, noch Etwas, man weiss nicht was, von ihm zurueckbleibe." Schelling, Ideen. etc., p. 18.

* Logic, vol. ii, p. 225. † *Ibid.*, p. 389.

The radical fallacy of the corpuscular as well as of the dynamical theory consists in the delusion that the conceptual elements of matter can be grasped as separate and real entities. The corpuscular theorists take the element of *inertia* and treat it as real by itself, while Boscovich, Faraday, and all those who define atoms or molecules as "centers of force," seek to realize the corresponding element, *force*, as an entity by itself. In both cases products of abstraction are mistaken for kinds of reality.

A satisfactory examination of the conceptual terms *inertia* and *force*, and of their true implications, is impossible here without anticipating considerations that properly belong to the following chapters. The essential correlation of inertia with force is evinced by its earliest definitions. Newton expressly speaks of inertia as of a force. "There is inherent," he says,* "in matter a force, a power of resistance, in virtue of which every body, as far as in it lies, perseveres in a state of rest or of uniform rectilinear motion." In the definition since Newton's time, this mode of expression has usually been discarded. Young † defines inertia as "the incapability of matter of altering the state into which it is put by any external cause, whether that state be rest or motion;" and, similarly, Whewell,‡ as "the quantity of *matter* considered as resisting the communication of motion." All these definitions imply, however, that the forces moving a body or a particle as a whole are strictly and absolutely extraneous forces. In the language of Newton,# force is "*impressed* upon a body, and exerted upon it to change its state of rest or uniform motion in a straight line."

* Princ., Def. iii.
† Mechanics, p. 117.
‡ Mechanics, p. 245.
\# Princ., Def. iv.

There is little difficulty in understanding how the disjunction of matter and force and the etymological import of the word "inertia" led to the assumption that matter is essentially passive, or, as it is commonly expressed, dead. When a body is considered by itself —conceptually detached from the relations which give rise to its attributes—it is indeed inert, and all its action comes from without. But this isolated existence of a body is a pure fiction of the intellect. Bodies exist solely in virtue of their relations; their reality lies in their mutual action. Inert matter, in the sense of the mechanical theory, is as unknown to experience as it is inconceivable in thought. Every particle of matter of which we have any knowledge attracts every other particle in conformity to the laws of gravitation; and every material element exerts chemical, electrical and other force upon other elements which, in respect of such force, are its correlates. A body can not, indeed, move itself; but this is true for the same reason that it can not exist in and by itself. The very presence of a body in space and time, as well as its motion, implies interaction with other bodies, and therefore *actio in distans;* consequently all attempts to reduce gravitation or chemical action to mere impact are aimless and absurd.

Physicists are perfectly aware that the sense commonly attached to the word *inertia* in its application to matter is spurious. "The incapacity of all material points," says M. Poisson, "to put themselves in motion, or to change the motion which has been communicated to them without the aid of a force, is what constitutes the inertia of matter. This word does not signify that matter is incapable of action; on the contrary, every material point at all times finds the principle of its movement in the action of other points, but never in itself." *

In spite of statements like this, however, and notwithstanding the clear apprehension, by leading physicists, of the true import of the doctrine of inertia, the phantom of "dead matter" incessantly obtrudes itself anew as the basis of cosmological speculations. Thus, Professor Philip Spiller, the author of a very serviceable manual of physics, and a prolific writer on scientific subjects, some years ago published a cosmological treatise,† whose theorems are founded on the express proposition that "no material constituent of a body, no atom, is in itself originally endowed with force, but that every such atom is absolutely dead, and without any inherent power to act at a distance." ‡ It appears from the further contents of this treatise that he not only denies force to the atoms taken singly, but that he also denies the possibility of their mutual action. He is driven, therefore, to the assertion of the independent substantiality of force; and, accordingly, he assumes force to be an all-pervading *quasi*-material presence—as he terms it, an incorporeal matter (unkoerperlicher Stoff). In utter disregard of the fundamental correlation of force and mass, Spiller identifies his force-substance with the omniferous æther, so that

* "L'impossibilité où sont tous les points matériels de se mettre en mouvement ou de changer le mouvement qui leur a été communiqué, sans le secours d'une force, est ce qu'on entend par *l'inertie* de la matière. Ce mot ne signifie pas que la matière soit incapable d'agir; car, au contraire, chaque point matériel trouve toujours dans l'action d'autres points matériels, mais jamais en lui même, le principe de son mouvement." Poisson, Traité de Mécanique, liv. ii, chap. i, 110.

† Der Weltaether als kosmische Kraft. Berlin, Denicke's Verlag, 1873.

‡ *Loc. cit.*, p. 4.

this hypostasized half-concept, which, in the view of all other physicists, is not only imponderable, but destitute of cohesive, chemical, thermal, electric, and magnetic forces (which, indeed, *must* be destitute of them if it is to stand as the mere substratum of these various modes of motion), and is, therefore, still more "dead," if possible, than ordinary matter, now suddenly, without changing its name, and without ceasing to be the substratum of luminar or other undulations, comes to be the very quintessence of all possible energy.

Professor Spiller's speculations are a strange revival of the well-known dreams of Kepler, who imagined that the planets were borne and carried along in their courses by an "immateriate species" (*species immateriata*) capable of overcoming the inertia of bodies.* Kepler's "immateriate species" is the same wooden iron which Spiller exhibits under the name "incorporeal matter," the only difference being that the absurdity of Kepler's chimera was less glaring in the hazy dawn of the mechanical notions of his time than the extravagance of Spiller's conceit in the light of the scientific atmosphere of our day.

What possible part Spiller's dead matter could perform in any cosmological scheme, it is difficult to see. Unchangeable particles destitute of gravity and all other force, even if the action of force upon them were conceivable, must be equally acted upon from all sides by the omnipresent æther, and could not, therefore, in

* "Relinquitur igitur, ut quemadmodum lux omnia terrena illustrans species est immateriata ignis illius, qui est in corpore Solis: ita virtus hæc, planetarum corpora complexa et vehens, sit species immateriata ejus virtutis, quæ in ipso Sole residet, inæstimabilis vigoris, adeoque actus primus omnis motûs mundani," etc. Kepler, De Motibus Stellæ Martis, pars tertia, cap. xxxiii; Kepleri Opp., ed. Frisch, vol. iii, p. 302.

any way help to establish differences of density, or other differences not contained in, or evolvable from, the æther itself. They could not even add to the extension of a body, much less to its hardness, being wholly without the power of resistance; but, waiving this, and granting that extension without resistance is possible, they would simply be bubbles of void space encysted in the universal æther, and to the differentiation of this æther alone all the phenomena of the material world would be due.

The prevailing errors respecting the inertia of matter have naturally led to corresponding delusions as to the nature of force. Here we are met, *in limine*, by an ambiguity in the meaning of the term force in physics and mechanics. When we speak of a "force of nature," we use the word force in a sense very different from that which it bears in mechanics. A "force of nature," is a survival of ontological speculation; in common phraseology the term stands for a distinct and real entity. But, as a determinate mechanical function, force is simply the rate of change of momentum—mathematically expressed, the differential of momentum at a given instant of time. "Momentum," says Mr. Tait,* "is the time-integral of force, because force is the rate of change of momentum." In the canonical text-books on physics, force is defined as the cause of motion. "Any cause," says Whewell,† "which moves or tends to move a body, or which changes or tends to change its motion, is called force." So Clerk Maxwell: ‡ "Force is whatever changes or tends to change the motion of a body by altering either its direction or its magnitude." Far greater insight into the nature of

* On Some Recent Advances in Physical Science, second ed., p. 347.

† Mechanics, p. 1.

‡ Théory of Heat, p. 83.

force is exhibited in the definition of Somoff, though the word "cause" is retained: "A material point is moved by the presence of matter without it. This action of extraneous matter is attributed to a cause which is named force." * Taking these definitions as correctly representing the received theories of physical science, it is manifest, irrespective of the considerations I have presented in this and the preceding chapters, that force is not an individual thing or entity that presents itself directly to observation or to thought, but that, so far as it is treated as a definite and unital term in the operations of thought, it is purely an incident to the conception of the interdependence of moving masses. The cause of motion, or of the change of motion, in a body is the condition or group of conditions upon which the motion depends; and this condition or group of conditions is always a corresponding motion, or change of motion, of the bodies outside of the body in question which are its dynamical correlates. † Otherwise expressed, force is a mere inference from the motion itself under the universal conditions of reality, and its measure and determination lie solely in the effect for which it is postulated as a cause; it has no other existence. The only reality of force and its action is the correspondence between physical phenomena in conformity with the principle of the essential relativity of all forms of physical existence.

That force has no independent reality is so plain and obvious that it has been proposed by some thinkers

* Somoff, Theoretische Mechanik (trans. by Ziwet), vol. ii, p. 155.

† "Der gegenwaertig klar entwickelte mechanische Begriff der Kraft," says Zoellner (Natur der Kometen, p. 323), "enthaelt nichts Anders als den Ausdruck einer raeumlichen und zeitlichen Beziehung zweier Koerper."

to abolish the term *force*, like the term *cause*, altogether. However desirable a sparing use of such terms may be (as is illustrated in the clearness of some modern mechanical treatises *), it is impracticable wholly to dispense with it, for the reason that the conceptual element force, when properly interpreted in terms of experience, is a legitimate incident to the conception of physical action, and, if its name were disused, it would instantly reappear under another name. There are few concepts which have not, in science as well as in metaphysics, given rise to the same confusion that prevails in regard to "force" and "cause;" and the blow leveled at these would demolish all concepts whatever. Nevertheless, it is of the greatest moment, in all speculations concerning the interdependence of physical phenomena, never to lose sight of the fact that force is a purely conceptual term, and that it is not a distinct tangible or intangible thing.

How imperfectly all this is apprehended in our time appears upon the most cursory examination of elementary treatises on physics as well as original scientific essays. The relation of force to mechanical motion is incessantly spoken of as a "fact ascertained by observation and verified by experiment." In an article published in July, 1872, it is said: "In regard to the first question (What produces motion?) there is no diversity of opinion. All agree that what produces change or causes motion is *force*." † The obvious meaning of this is, that it might possibly admit of question whether material change or motion is produced by force or by

* Cf. e. g. Kirchhoff, Vorlesungen ueber mathematische Physik. Heidelberg, 1876.

† What determines Molecular Motion, etc. By James Croll. Phil. Mag., fourth series, vol. xl, p. 37.

something else, and that physicists, on the whole, have come to the conclusion that it is produced by force. Such a question ought, indeed, to be gravely pondered! It is like the question which Mr. Sachs, in his despair, propounded to the world: "Who will assure us that the star which the astronomers regard as Uranus is Uranus in fact?" *

Physicists generally, however, are in still greater confusion as to the nature of force in another respect. Bodies are said to be *endowed* with a definite quantity of force; it is assumed that to every particular body or atom belongs, or that in such body or atom is inherent, an invariable measure of energy. This statement, besides involving the conceit just noted of the independent reality of force, implies the assumption that force can be an attribute or concomitant of a single particle as such, ignoring the fact, which is otherwise well known to physicists, that the very conception of force depends upon the relation between two terms at least. "Force," says Clerk Maxwell,† "is but one aspect of that mutual action between two bodies which is called by Newton Action and Reaction, and which is now more briefly expressed by the single word Stress." And in another place: ‡ "If we take into account the whole phenomenon of the action between two portions of matter, we call it Stress. . . . But, if we confine our attention to one of the portions of matter, we see, as it were, only one side of the transaction—namely, that which affects the portion of matter under our consideration—and we call this aspect of the phenomenon, with reference to its effect, an External Force acting on that

* Das Sonnensystem, oder neue Theorie vom Bau der Welten, von S. Sachs, p. 193 (quoted by Fechner).

† Matter and Motion, ci.

‡ *Ib.*, xxxvi, xxxviii.

portion of matter, and with reference to its cause we call it the Action of the other portion of matter. The opposite aspect of the stress is called the Reaction on the other portion of matter." To the same effect is the statement of Rankine:* "Force is an action between two bodies either causing or tending to cause change in their relative rest or motion." It follows that a "constant central force," as belonging to an individual atom or molecule in and by itself, is an impossibility.

* Applied Mechanics, fourth ed., p. 15.

CHAPTER XI.

CHARACTER AND ORIGIN OF THE MECHANICAL THEORY (CONTINUED).—ITS EXEMPLIFICATION OF THE THIRD RADICAL ERROR OF METAPHYSICS.

THERE are few beliefs which are generally held to be more indubitable than that of the absolute solidity of matter. With the exception of Descartes and his immediate followers, whose assertion that matter is nothing but extension is clearly indefensible, philosophers and physicists alike have always placed solidity and impenetrability in the front rank of its primary qualities. And this belief, in view of the observed transformations of material things, unavoidably leads to the doctrine that matter consists of indivisible, absolutely rigid particles. The opinion of Tyndall, expressed in the passage quoted in the last chapter from his Liverpool address, is the opinion of the bulk of scientific men, as well as of persons without scientific training. To all of them, as to Tyndall, it seems absurd to deny that the conception of matter necessarily involves the notion of definite, tangible, and indestructible solidity. It is the general tacit assumption that, of the three molecular states, or states of aggregation, in which matter presents itself to the senses—the solid, the liquid, and the gaseous—the last two are simply disguises or complications of the first; that a gas, for instance, is in fact a group or cluster of solids, like a cloud of dust, differing from such a cloud only by the greater regularity in the forms and distances of the particles whereof it is composed, and by the fact that these particles are controlled in the case of a gas by their mutual attractions and repulsions, while in the case of the cloud of dust they are under the sway of extrinsic forces. And, while the transition of the three molecular states into each other in regular and invariable order is too obvious to be ignored, it is supposed that the solid is the primary state of which the liquid and gaseous, or aëriform, states are simply derivatives, and that, if these states are considered as evolved the one from the other, the order of evolution is from the solid to the vapor or gas. In this view the solid form of matter is not only the basis and origin of all its further determinations—of all its evolutions and changes—but it is also the true and typical element of its mental representation and conception.

While this view of the relation between the molecular states of matter is universally prevalent, it is not difficult to show that it is inconsistent with the facts. All evolution proceeds from the relatively Indeterminate to the relatively Determinate, and from the comparatively Simple to the comparatively Complex. And (confining our attention, for the moment, to the two extreme terms of the evolution, the solid and the gas, and ignoring the intermediate liquid) a comparison of the gaseous with the solid state of matter at once shows that the former is, not the end, but the beginning of the evolution. The gas is not only comparatively indeterminate—without fixity of volume, without crystalline or other structure—but it also exhibits, in its functional manifestations, that simplicity and regularity which are characteristic of all types or primary forms.

Looking, *first*, to the purely physical aspect of a gas—I speak, of course, only of gases which are approximately perfect, to the exclusion of vapors at low temperature and of gases which are readily coercible: its volume expands and contracts as the pressure to which it is subjected; its velocity of diffusion is inversely proportional to the square root of its density; its rate of expansion is uniform for equal increments of temperature; its specific heat is the same at all temperatures, and, in a given weight, for all densities and under all pressures; the specific heats of equal volumes of simple and incondensible gases, as well as of compound gases formed without condensation, are the same for all gases of whatever nature, and so on. In all these respects the contrast with the liquid as well as the solid form, the relations of whose volumes, or structures, or both, to temperature and to mechanical pressure or other force are complicated in the extreme, is great and striking. But this contrast becomes still more signal, secondly, under the chemical aspect. We can not, in any proper sense, assign the proportions of volume in which the combination of solids and liquids takes place—indeed, the combination of solids as such is impossible—and the numbers expressive of the proportions of the combining weights upon their face exhibit an appearance of irrelation and irregularity which the most sustained endeavors of scientific men (such as Dumas, Stas, H. Carey Lea, Cooke, L. Meyer, Mendelejeff, Baumhauer) have been unable to obliterate. In the combination of gases, on the contrary, all is simplicity and order. "The ratio of volumes in which gases combine is always simple, and the volume of the resulting gaseous product bears a simple ratio to the volumes of its constituents"—such is the law of Gay-Lussac. By weight, the ratio of combination between hydrogen and chlorine is 1 to 35·5; by volumes, one volume of hydrogen combines with one volume of chlorine (the volumes being taken, of course, at the same pressures and temperatures) so as to form two volumes of hydrochloric acid. Oxygen and hydrogen combine in the proportion of 16 to 2 by weight; but one volume of oxygen combines with two volumes of hydrogen, forming two volumes of watery vapor. Nitrogen and hydrogen, whose atomic weights, so called, are 14 and 1 respectively, combine in the simple ratio of one volume of nitrogen to three volumes of hydrogen, the combination resulting in two volumes of gaseous ammonia. And carbon, whose "atomic weight" is 12, though it can not be actually obtained in gaseous form, is assumed by all chemists (for reasons not necessary to state here) to combine with hydrogen in the ratio of one volume to four, so as to yield two volumes of marsh-gas.

All this warrants the conclusion that if there be a typical and primary state of matter, it is, not the solid, but the gas. And, this being so, it follows that the molecular evolution of matter conforms to the law of all evolution in proceeding from the indeterminate to the determinate, from the simple to the complex, from the gaseous to the solid form. Inasmuch, therefore, as the explanation of any phenomenon aims at the exhibition of its genesis from its simplest beginnings, or from its earliest forms, the gaseous form of matter is the true basis for the explanation of the solid form, and not, conversely, the solid for the explanation of the gas.

From the foregoing considerations, I take it to be clear that the true relation between the molecular states of matter is the exact reverse of that universally assumed. The universality of this assumption, however, indicates that it is due, not to a mere chance error of reasoning, but to some natural bias of the mind. The question arises, therefore: What is the origin of this prevalent delusion respecting the constitution of matter? I believe the answer to this question to be exceedingly simple, and important in proportion to its simplicity. One of the fallacies to which the human intellect is liable by reason of the laws of its growth, and which I have ventured to call *structural* fallacies, is that the intellect tends to confound the order of the genesis of its ideas respecting material objects with the order of the genesis of these objects themselves. I have heretofore shown that the progress of our knowledge depends upon analogy—upon a reduction of the Strange and Unknown to the terms of the Familiar and Known. In a certain sense it is true, what has been often said, that all cognition is recognition. "Man constantly institutes comparisons," says Pott,* "between the new which presents itself to him and the old which he already knows." That this is so is shown by the development of language. The great agent in the evolution of language is metaphor—the transference of a word from its ordinary and received meaning to an analogous one. This transference of the name descriptive of a known and familiar thing to the designation of an unknown and unfamiliar thing typifies the proceeding of the intellect in all cases where it deals with new and strange phenomena. It assimilates these phenomena to those which are known; it identifies the Strange, as far as possible, with the Familiar; it apprehends that which is extraordinary and uncommon in terms of that which is ordinary and common. But that which is most obvious

* Etymologische Forschungen, 2d ed., vol. ii, p. 139.

to the senses is both the earliest and most persistent presence in consciousness, and thus receives the stamp of the greatest familiarity. Now, the most obtrusive form of matter is the solid, and for this reason it is that form which is first cognized by the infant intellect of mankind, and thus serves as the basis for the subsequent recognition of other forms. Accordingly we find that, on the early stages of human history, the solid alone was apprehended as material. It was long before even atmospheric air, obtrusive as it was in wind and storm, came to be known as a form of matter. To this day words signifying wind or breath—animus, spiritus, Geist, ghost, etc.—are the terms denoting that which is the fundamental correlate of matter, even in the languages of civilized nations. And it is very questionable whether either the ancient philosophers or the mediæval alchemists distinctly apprehended any aëriform substance, other than atmospheric air, as material. It is certain that up to the time of Van Helmont, in the latter part of the sixteenth and the first decades of the seventeenth century, aëriform matter was not the subject of sustained scientific investigation.

It is obvious then, that, while the progress of evolution in nature is from the aëriform to the solid state of matter, the progress of the evolution of knowledge in the minds of men was, conversely, from the solid to the aëriform; and, as a consequence, the aëriform or gaseous state came to be apprehended as a mere modification of solidity. For the same reason, the first form of material action which was apprehended by the dawning intellect of man was the interaction between solids —mechanical interaction—and from this, again, it followed that the difference between the solid and the gas was apprehended as a mere difference of distance between the solid particles, as produced by mechanical motion.

Again: familiarity, in the minds of ordinary men, is universally confounded with simplicity. And, the explanation of a phenomenon aiming, as we have seen, at an exhibition of its genesis from its simplest beginnings, the mind, in its attempts to explain the gaseous form, naturally retraces the steps in the evolution of its ideas concerning matter—of its concepts of matter—back to the earliest, most familiar, and therefore apparently simplest form in which matter was and is apprehended, and assumes the solid particle, the atom, as the ultimate fact, as the primary element for all representation and conception of material existence.

The assumption of the identity of the order of conception with the order of reality (the third of the fallacious assumptions enumerated in the ninth chapter) is one of the most fatal errors of ontological speculation, and has been signalized as such by J. S. Mill, who fails, however, to discover the true source of this error as pointed out above, attributing (as usual) the order and connection of our ideas to mere fortuitous association. "A large proportion of the erroneous thinking," he says,* "which exists in the world, proceeds on the tacit assumption that the same order must obtain among the objects of nature which obtains among our ideas of them." The inveteracy of this assumption and its irrepressible dominance in ontological speculation might be shown by numerous examples. Spinoza makes the distinct declaration that "the order and connection of ideas are the same as the order and connection of things." †

* Logic, 8th ed., p. 521.

† "*Ordo et connexio idearum idem est ac ordo et connexio rerum.*" Eth. ii, prop. 7.

And even in a late treatise on logic we read that "the logical catenation of ideas corresponds to the real catenation of things." * Here again, then, the metaphysical character of the atomo-mechanical theory becomes manifest.

Although the opinion that solidity and impenetrability are not only indispensable, but also perfectly simple attributes of matter, is all but universal, there are some thinkers who do not fail to see that it is due to a prejudice of the intellect. "In the hypothesis," says M. Cournot, † "to which modern physicists have been led—that of atoms kept at a distance from each other, and even at distances which, though inappreciable in experience, are nevertheless very great in comparison with the dimensions of the atoms or elementary corpuscles—there is nothing that compels the conception of atoms as hard or solid little bodies rather than as small, soft, flexible, or liquid masses. The preference which we give to hardness over softness, the tendency to represent the atom or primordial molecule as a miniature of a solid body, rather than as a fluid mass of the same size, are therefore nothing but prejudices of education resulting from our habits and the conditions of our animal life. Consequently there is nothing more unfounded than the old belief—so deeply rooted in the old scholastics and perpetuated even in modern doctrines—which makes impenetrability added to extension the fundamental property of matter and of bodies. It is too clear that atoms which could never come in contact could much less penetrate each other, so that the quality said to be fundamental would, on the contrary, be a

* "*L'enchainement logique des idées correspond à l'enchainement réel des choses.*" Delbœuf, Logique, p. 91.

† De l'Enchainement, etc., vol. i, p. 246 *seq.*

useless, idle quality which would never come into play and would never be part of the explanation of any phenomenon, and the assertion of whose existence would be gratuitous. The same thing is to be said of extension as an attribute or quality of atoms, inasmuch as, upon last analysis and in the present state of the sciences, all the explanations that can be given of physico-chemical phenomena are perfectly independent of the hypotheses which can be framed respecting the figures and dimensions of elementary atoms or molecules. As to bodies of finite dimensions falling under our senses, they are all certainly penetrable ; and so far as they are concerned the continuity of forms of extension is but an illusion.

"In the bodies that fall under our senses, solidity and rigidity, like flexibility, softness, or fluidity, are so many very complex phenomena which we attempt to explain, as best we can, by the aid of hypotheses respecting the law of the forces that maintain the elementary molecules at (definite) distances, and respecting the extent of their sphere of action, as compared with the number of molecules embraced in that sphere and with the distances between them. Now, while the familiar notion of bodies in the solid state has suggested the conception of the rigid corpuscle or elementary atom as a philosophical and scientific principle of explanation, there is nothing more difficult to explain satisfactorily, by means of the conception of atoms, than the constitution of bodies in the state of solidity."

I have already cited, in the seventh chapter, a passage of similar import from the lectures of M. Cauchy, in which that distinguished mathematician questions the necessity of attributing to matter either impenetrability or extension (without which, or either of which, there can, of course, be no solidity) as a primary quality.

Solidity, in the sense in which it is attributed to the atom, is not a fact, but the hypostasis of an abstraction. As M. Cournot observes, an absolutely solid body is unknown to experience. The consistency of the bodies which present themselves to the experimental physicist depends upon the preponderance or balance of forces, such as the forces of cohesion, crystallization, and heat; and the assumption of the absolute solidity of matter results from that superficial and imperfect apprehension of the data of sense (in conjunction with the disregard of the essential relativity of all the properties of things to be considered more at length hereafter) which is reflected in all the early notions of mankind.

The same primitive, perfunctory and incomplete apprehension of the data of sense has given rise to the further assumption that all physical action is by impact. The only interaction between bodies that is directly appreciable by the senses of sight and touch is a change in the state of rest or motion by collision. A thrust is, therefore, the earliest and most familiar of all the observable actions of one body upon another. And when impact takes place between two solids moving with different velocities, or (what is the same thing) between a solid in motion and another solid at rest, the ordinary observer sees nothing more than a displacement of one body by the other, and a direct transference of motion. This displacement and transference are taken to be instant and the bodies are supposed to be absolutely rigid. But this observation of the fact is as crude as its interpretation is inaccurate. A more careful study of the phenomena shows that there is no such immediate displacement; that there is no direct transference of motion; that the bodies are not absolutely rigid; that the apparently simple impact of solids is a very complex series or group of incidents involving not only direct action and reaction, but also alternate compression and expansion, a loosening and tightening of cohesive and crystalline bonds, transformations of rectilinear into vibratory, of molar into molecular motion, evolution and involution of energy—in short, momentary, if not permanent changes in all, or nearly all, the properties of bodies between which the impact occurs. In view of this: what does the demand of the atomo-mechanical theory, to admit no interaction between bodies other than that of impact, imply? Nothing less than this, that the first rudimentary and unreasoned impressions of the untutored savage shall stand for ever as the basis of all possible science.

Suppose that Hobbes had been familiar with the incidents to the origin and transformation of motion, as they have been brought to light by observation and experiment in recent times; suppose he had been able, as clearly as Helmholtz and Mayer, or Thomson and Joule, to trace, not only the rotatory as well as revolutionary motions of our planet, but also every disturbance upon it—every blow dealt by a living hand and every shock caused by the fall or projection of an inanimate mass—to the undifferentiated energy of a primordial gaseous spheroid from which the sun and the earth are supposed to have been slowly precipitated or evolved; suppose that, whenever he observed the phenomenon of impact between two solids and the apparent transference of visible motion from one to the other, his thoughts had involuntarily run back to the embryo form of this phenomenon, the alternate contraction and expansion of a formless mobile gas: would he have written the sentence that "there can be no cause of motion except in a body contiguous and moved"?

The logical and mathematical inadmissibility of the assumption of the absolute solidity of extended atoms or molecules was pointed out in the early part of the last century by John Bernoulli, who showed that it involved the conception of an infinite power of resistance to deformation or compression. And that solidity is not the simplest, but the most complicated phase of material consistency, was urged, nearly sixty years ago, by Fries, who objected to all atomic theories that "they assumed that which is the most difficult, viz.: the constitution of definite forms, as an original datum and as the starting-point of explanation,"* whereas "the great difficulty of the mathematical philosophy of nature is the possibility of rigid bodies." †

The absolute solidity of matter is one of the forms in which the pseudo-concept of "being *per se*" or "simple existence" is hypostasized, in disregard of the essential relativity of material things, which I propose to discuss in the next chapter.

* Fries, Mathematische Naturphilosophie (Heidelberg, 1822), p. 446.

† *Id. ib.*, p. 616. It will be noticed that Fries here anticipates the observation of Cournot heretofore cited.

CHAPTER XII.

CHARACTER AND ORIGIN OF THE MECHANICAL THEORY (CONTINUED).—ITS EXEMPLIFICATION OF THE FOURTH RADICAL ERROR OF METAPHYSICS.

THE reality of all things which are, or can be, objects of cognition, is founded upon, or, rather, consists in, their mutual relations. A thing in and by itself can be neither apprehended nor conceived; its existence is no more a presentation of sense than a deliverance of thought. Things are known to us solely through their properties; and the properties of things are nothing else than their interactions and mutual relations. "Every property or quality of a thing," says Helmholtz * (speaking of the inveterate prejudice according to which the qualities of things must be analogous to, or identical with, our perceptions of them), "is in reality nothing but its capability of producing certain effects on other things. The effect occurs either between like parts of the same body so as to produce differences of aggregation, or it proceeds from one body to another, as in the case of chemical reactions; or the effects are upon our organs of sense and manifest themselves as sensations such as those with which we are here concerned (the sensations of sight). Such an effect we call a 'property,' its reagent being understood

* Die neueren Fortschritte in der Theorie des Sehens. Pop. wiss. Vortraege, ii, 55 *seq.*

without being expressly mentioned. Thus we speak of the 'solubility' of a substance, meaning its behavior toward water; we speak of its 'weight,' meaning its attraction to the earth; and we may justly call a substance 'blue' under the tacit assumption that we are only speaking of its action upon a normal eye. But, if what we call a property always implies a relation between two things, then a property or quality can never depend upon the nature of one agent alone, but exists only in relation to and dependence on the nature of some second object acted upon. Hence, there is really no sense in talking of properties of light which belong to it absolutely, independently of all other objects, and which are supposed to be representable in the sensations of the human eye. The notion of such properties is a contradiction in itself. They can not possibly exist, and therefore we can not expect to find any coincidence of our sensations of color with qualities of light."

The truth which underlies these sentences is of such transcendent importance that it is hardly possible to be too emphatic in its statement, or too profuse in its illustration. The real existence of things is coextensive with their qualitative and quantitative determinations. And both are in their nature relations, quality resulting from mutual action, and quantity being simply a ratio between terms neither of which is absolute. Every objectively real thing is thus a term in numberless series of mutual implications, and forms of reality beyond these implications are as unknown to experience as to thought. There is no absolute material quality, no absolute material substance, no absolute physical unit, no absolutely simple physical entity, no absolute physical constant, no absolute standard, either of quantity or quality, no absolute motion, no absolute rest, no absolute time, no absolute space. There is no form of material existence which is either its own support or its own measure, and which abides, either quantitatively or qualitatively, otherwise than in perpetual change, in an unceasing flow of mutations. An object is large only as compared with another which, as a term of this comparison, is small, but which, in comparison with a third object, may be indefinitely large; and the comparison which determines the magnitude of objects is between its terms alone, and not between any or all of its terms and an absolute standard. An object is hard as compared with another which is soft, but which, in turn, may be contrasted with a third still softer; and, again, there is no standard object which is either absolutely hard or absolutely soft. A body is simple as compared with the compound into which it enters as a constituent; but there is and can be no physically real thing which is absolutely simple.*

It may be observed, in this connection, that not only the law of causality, the conservation of energy, and the indestructibility of matter, so called, have their root in the relativity of all objective reality—being, indeed, simply different aspects of this relativity—but that New-

* One of the most noteworthy specimens of ontological reasoning is the argument which infers the existence of absolutely simple substances from the existence of compound substances. Leibnitz places this argument at the head of his "Monadology." "*Necesse est*," he says, "*dari substantias simplices quia dantur composita; neque enim compositum est nisi aggregatum simplicium.*" (Leibnitii, Opera omnia, ed. Dutens, t. ii, p. 21.) But the enthymeme is obviously a vicious paralogism—a fallacy of the class known in logic as fallacies of suppressed relative. The existence of a compound substance certainly proves the existence of component parts which, *relatively to this substance*, are simple. But it proves nothing whatever as to the simplicity of these parts in themselves.

ton's first and third laws of motion, as well as all laws of least action in mechanics (including Gauss's law of movement under least constraint), are but corollaries from the same principle. And the fact that everything is, in its manifest existence, but a group of relations and reactions at once accounts for Nature's inherent teleology.

Although the truth that all our knowledge of objective reality depends upon the establishment or recognition of relations is sufficiently evident and has been often proclaimed, it has thus far been almost wholly ignored by men of science as well as by metaphysicians. It is to this day assumed by physicists and mathematicians, no less than by ontologists, that all reality is in its last elements absolute. And this assumption is all the more strenuously insisted on by those whose scientific creed begins with the proposition that all our knowledge of physical things is derived from experience. Thus the mathematician, who fully recognizes the validity of this proposition and at the same time concedes that we have, and can have, no actual knowledge of bodies at rest or in motion, except in relation to other bodies, nevertheless declares that rest and motion are real only in so far as they and their elements, space and time, are absolute. The physicist reminds us at every step that in the field of his investigations there are no *a priori* truths and that nothing is known of the world of matter save what has been ascertained by observation and experiment; he then announces as the uniform result of his observations and experiments, that all forms of material existence are complex and variable; and yet he avers that not merely the laws of their variation are constant, but that the real constituents of the material world are absolutely simple, invariable, individual things.

The assumption that all physical reality is in its last elements absolute—that the material universe is an aggregate of absolutely constant physical units which in themselves are absolutely at rest, but whose motion, however induced, is measurable in terms of absolute space and absolute time—is obviously the true logical basis of the atomo-mechanical theory. And this assumption is identical with that which lies at the root of all metaphysical systems, with the single difference that in some of these systems the physical substratum of motion (termed the "substance" of things) is not specialized into individual atoms.

To show how irrepressibly the ontological prejudice, that nothing is physically real which is not absolute, has asserted itself in science during the last three centuries, I propose briefly to review the doctrines of some of the most eminent mathematicians and physicists respecting space and motion (and, incidentally, time), beginning with those of Descartes.

In the introductory parts of his Principia, Descartes states in the most explicit terms that space and motion are essentially relative. "In order that the place [of a body] may be determined," he says,* "we must refer to other bodies which we may regard as immovable, and accordingly as we refer to different bodies it can be said that the same thing does, and does not, change its place. Thus, when a ship is carried along at sea, he who sits at the stern remains always at the same place in reference to the parts of the ship among which he retains the same position; but he continually changes his place in reference to the shores. . . . And besides, if we allow that the earth moves and proceeds—precisely as far from west to

* Princ. ii, § 18.

east as the ship meanwhile is carried from east to west—we shall say again that he who sits at the stern does not move his place, because we determine it with reference to some immovable points in the heavens. But, if finally we concede that no truly immovable points are to be found in the universe, as I shall hereafter show is probable, our conclusion will be that there is nothing which has a fixed place except so far as it is determined in thought." *

Statements to the same effect are found in various other parts of the same book.† And of space Descartes does not hesitate to say that is really nothing in itself, and that "void space" is a contradiction in terms—that, as Sir John Herschel puts it,‡ "if it were not for the foot-rule between them, the two ends of it would be in the same place." But, in the further progress of his discussions, having meanwhile declared that God always conserves in the universe the same quantity of motion, he all at once takes it for granted # that motion and space are absolute and therefore real entities.

This inconsistency of Descartes is severely censured by Leibnitz. "It follows," says Leibnitz,‖ "that motion is nothing but a change of place, and thus, so far

* The illustration of the relativity of motion by the motion of a ship is of constant recurrence whenever reference is had to the question discussed in the text. Cf. Leibnitz, Opp. ed. Erdmann, p. 604; Newton, Princ., Def. viii, Schol. 3; Euler, Theoria Motûs Corporum Solidorum, vol. i, 9, 10; Berkeley, Principles of Human Knowledge, § 114; Kant, Metaphysische Aufansgruende der Naturwissenschaft, Phor. Grundsatz I; Cournot, De l'Enchainement, etc., vol. i, p. 56; Herbert Spencer, First Principles, chapter iii, § 17, etc., etc.

† E. g., Princ., ii, 24, 25, 29, etc.

‡ Familiar Lectures, p. 455.

Princ., ii, §§ 37-39.

‖ Leibn., Opp. math., ed. Gerhardt, sect. ii, vol. ii, p. 247.

as phenomena are concerned, consists in a mere relation. This Cartesius also acknowledged; but in deducing his consequences he forgot his own definition and framed his laws of motion *as though motion were something real and absolute*." As will be noticed, Leibnitz here assumes, as a matter of course, that what is real is also absolute. In view of this it is hardly surprising that he, too, falls into the same inconsistency with which he charges Descartes, and, in his letters to Clarke, speaks of "absolutely immovable space" and an "absolutely veritable motion of bodies." *

Newton, in the great Scholium to the last of the "Definitions" prefixed to his Principia, sharply distinguishes between absolute and relative time and motion. "Absolute and mathematical time," he says,† "in itself and in its nature without relation to anything external, flows equally and is otherwise called duration; relative, apparent and vulgar time is any sensible and extrinsic, accurate or unequal measure of duration by motion which is ordinarily taken for true time. . . . Absolute is distinguished from relative time in astronomy by the equation of vulgar time. For the natural days, which are vulgarly taken in the measurement of time as equal, are unequal. . . . *It may be that there is no equable motion by which time is accurately measured*." ‡

"Absolute space, in its nature without relation to anything external, always remains similar and immovable; of this (absolute space) relative space is any movable measure or dimension which is sensibly defined by its place in reference to bodies, and is vulgarly taken

* Opp. ed. Erdmann, pp. 766, 770.

† Princ. (ed. Le Seur & Jacq.), p. 8.

‡ *L. c.*, p. 10.

for immovable space.* . . . We define all places by the distances of things from some [given] body which we take as immovable. . . . *It may be that there is no body truly at rest to which places and motions are to be referred.*" †

Absolute motion, according to Newton, is "the translation of a body from one absolute place to another," and relative motion "the translation of a body from one relative place to another. . . . Absolute rest and motion are distinguished from relative rest and motion by their properties and by their causes and effects. It is the property of rest that bodies truly at rest are at rest in respect to each other. Hence, while it is possible that in the regions of the fixed stars, or far beyond them, there is some body absolutely at rest, it is nevertheless impossible to know from the relative places of bodies in our regions, whether any such distant body persists in the given position, and therefore true rest can not be defined from the mutual position of these" [i. e., the bodies in our regions]. . . . "It is the property of motion that the parts which retain their given positions to the wholes participate in their motion. For all the parts of rotating bodies tend to recede from the axis of motion, and the impetus of the moving bodies arises from the impetus of the parts. Hence, when the surrounding bodies move, those which move within them are relatively at rest. *And for this reason true and absolute motion can not be defined by their translation from the vicinity of bodies which are looked upon as being at rest.*‡ . . . The causes by which true and relative motions are distinguished from each other are the forces impressed upon bodies for the generation of motion. True motion is generated or

* *L. c.*, p. 9. † *Ib.*, p. 10. ‡ *Ib.*, pp. 10, 11.

changed solely by the forces impressed upon the body moved; but relative motion may be generated and changed without the action of forces upon it. For it is sufficient that forces are impressed upon other bodies to which reference is had, so that by their giving way a change is effected in the relation in which the relative motion or rest of the body consists.* . . . The effects by which absolute and relative motion are mutually distinguished are the forces by which bodies recede from the axis of circular motion. For in purely relative circular motion these forces are null, while in true and absolute motion they are greater or less according to the quantity of motion." †

It is apparent that in all these definitions Newton, like Descartes and Leibnitz, assumes real motion to be absolute, and that he takes the terms *relative motion* and *apparent motion* to be strictly synonymous, notwithstanding his express admission (in the passages which I have italicized) that in fact there may be neither absolute time nor absolute space. That admission naturally leads to the further admission that there may in fact be no absolute motion; but from this Newton recoils, resorting to the expedient of trying to find tenable ground for the distinction between absolute and relative motion, despite the possible non-existence of absolute time and space, in what he calls their respective causes and effects. But these causes and effects serve to distinguish, not relative from absolute change of position, but simply change of position in one body with reference to another from simultaneous changes of position in both with reference to a third.

Newton's doctrine is pushed to its last consequences

* *L. c.*, p. 11. † *Ib.*

by Leonhard Euler. In the first chapter of his "Theory of the Motion of Solid or Rigid Bodies," Euler begins with the emphatic declaration that rest and motion, so far as they are known to sensible experience, are purely relative. After referring to the typical case of the navigator in his ship, he proceeds:* "The notion of rest here spoken of, therefore, is one of relations, inasmuch as it is not derived solely from the condition of the point O to which it is attributed, but from a comparison with some other body A . . . . And hence it appears at once that the same body which is at rest with respect to the body A is in various motion with respect to other bodies. . . . What has been said of relative rest may be readily applied to relative motion; for when a point O retains its place with respect to a body A, it is said to be relatively at rest, and, when it continually changes that place, it is said to be relatively in motion.† . . . *Therefore motion and rest are distinguished merely in name and are not opposed to each other in fact, inasmuch as both may at the same time be attributed to the same point, accordingly as it is referred to different bodies. Nor does motion differ from rest otherwise than as one motion differs from another.*" ‡

After thus insisting upon the essential relativity of rest and motion, Euler proceeds, in the second chapter, "On the Internal Principles of Motion," to consider the question whether or not rest and motion are predicable of a body without reference to other bodies. To this question he unhesitatingly gives an affirmative answer, holding it to be axiomatic that "every body, even without respect to other bodies, is either at rest or in

* Theoria motûs Corp. Sol, etc., cap. i, explic. 2.

† *Ib.*, p. 7. ‡ *Ib.*, p. 8.

motion, i. e., is either absolutely at rest or absolutely in motion.* . . . "Thus far," he explains, "following the senses, we have not recognized any other motion or rest than that with respect to other bodies, whence we have called both motion and rest relative. But, if we now mentally take away all bodies but one, and if thus the relation by which we have hitherto distinguished its rest and motion is withdrawn, it will first be asked whether or not the conclusion respecting the rest or motion of the remaining body still stands. For, if this conclusion can be drawn only from a comparison of the place of the body in question with that of other bodies, it follows that, when these bodies are gone, the conclusion must go with them. *But, albeit we do not know of the rest or motion of a body except from its relation to other bodies, it is nevertheless not to be concluded that these things (rest and motion) are nothing in themselves but a mere relation established by the intellect, and that there is nothing inherent in the bodies themselves which corresponds to our ideas of rest and motion.* For, although we are unable to know quantity otherwise than by comparison, yet, when the things with which we instituted the comparison are gone, there is still left in the body the *fundamentum quantitatis*, as it were; for, if it were extended or contracted, such extension or contraction would have to be taken as a true change. Thus, if but one body existed, we should have to say that it was either in motion or at rest, inasmuch as it could not be taken as being both or neither. *Whence I conclude that rest and motion are not merely ideal things, born from comparison alone, so that there would*

* *Omne corpus, etiam sine respectu ad alia corpora, vel quiescit vel movetur, hoc est, vel absolute quiescit, vel absolute movetur.*" *Ib.*, p. 30 (cap. ii, axioma 7).

be nothing inherent in the body corresponding to them, but that it may be justly asked in respect to a solitary body whether it is in motion or at rest. . . . Inasmuch, therefore, as we can justly ask respecting a single body itself, without reference to other bodies, or under the supposition that they are annihilated, whether it is at rest or in motion, we must necessarily take one or the other alternative. But what this rest or motion will be, in view of the fact that there is here no change of place with respect to other bodies, we can not even think without admitting an absolute space in which our body occupies some given space whence it can pass to other places." * Accordingly Euler most strenuously insists on the necessity of postulating an absolute, immovable space. "Whoever denies absolute space," he says, "falls into the gravest perplexities. Since he is constrained to reject absolute rest and motion as empty sounds without sense, he is not only constrained also to reject the laws of motion, but to affirm that there are no laws of motion. For, if the question which has brought us to this point, What will be the condition of a solitary body detached from its connection with other bodies? is absurd, then those things also which are induced in this body by the action of others become uncertain and indeterminable, and thus everything will have to be taken as happening fortuitously and without any reason." †

That the basis of all this reasoning is purely ontological is plain. And, when the thinkers of the eighteenth century became alive to the fallacies of ontological speculation, the unsoundness of Euler's "axiom," that rest and motion are substantial attributive entities independent of all relation, could hardly escape their

* Theoria motus, etc., p. 31. † *Ib.*, p. 32.

notice. Nevertheless, they were unable to emancipate themselves wholly from Euler's ontological prepossessions. They did not at once avoid his dilemma by repudiating it as unfounded—by denying that motion and rest can not be real without being absolute—but they attempted to reconcile the absolute reality of rest and motion with their phenomenal relativity by postulating an absolutely quiescent point or center in space to which the positions of all bodies could be referred. Foremost among those who made this attempt was Kant.* In the seventh chapter of his "Natural His-

* It is remarkable how many of the scientific discoveries, speculations and fancies of the present day are anticipated or at least foreshadowed in the writings of Kant. Some of them are enumerated by Zoellner (Natur der Kometen, p. 455 *seq.*)—among them the constitution and motion of the system of fixed stars; the nebular origin of planetary and stellar systems; the origin, constitution and rotation of Saturn's rings and the conditions of their stability; the non-coincidence of the moon's center of gravity with her center of figure; the physical constitution of the comets; the retarding effect of the tides upon the rotation of the earth; the theory of the winds, and Dove's law. Fritz Schultze has shown (Kant and Darwin, Jena, 1875) that Kant was one of the precursors of Darwin. In this connection it is curious to note a coincidence (no doubt wholly accidental) in the example resorted to both by Kant and A. R. Wallace for the purpose of illustrating "adaptation by general law." The case put by both is that of the channel of a river which, in the view of the teleologists, as Wallace says (Contributions to the Theory of Natural Selection, p. 276 *seq.*), "must have been designed, it answers its purpose so effectually," or, as Kant expresses it, must have been scooped out by God himself. ("Wenn man die physisch-theologischen Verfasser hoert, so wird man dahin gebracht, sich vorzustellen, ihre Lanfrinnen waeren alle von Gott ausgehoehlt." Beweisgrund zu einer Demonstration des Dasein's Gottes, Kant's Werke, i, p. 232.) Even of the vagaries of modern transcendental geometry there are suggestions in Kant's essays, Von der wahren Schaetzung der lebendigen Kraefte, Werke v, p. 5, and Von dem ersten Grunde des Unterschiedes der Gegenden im Raume, *ib.*, p. 293—a fact which is not likely to conduce to the edification of those who, like J. K. Becker, Tobias, Weissenborn, Krause, etc., have raised the Kantian standard in defense of Eu-

tory of the Heavens"—the same work in which, nearly fifty years before Laplace, he gave the first outlines of the Nebular Hypothesis—he sought to show that in the universe there is somewhere a great central body whose center of gravity is the cardinal point of reference for the motions of all bodies whatever. "If in the immeasurable space," he says,* "wherein all the suns of the milky way have been formed, a point is assumed round which, from whatever cause, the first formative action of nature had its play, then at that point a body of the largest mass and of the greatest attractions, must have been formed. This body must have become able to compel all systems which were in process of formation in the enormous surrounding sphere to gravitate toward it as their center, so as to constitute an entire system, similar to the solar and planetary system which was evolved on a small scale out of elementary matter."

A suggestion similar to that of Kant has recently been made by Professor C. Neumann, who enforces the necessity of assuming the existence, at a definite and permanent point in space, of an absolutely rigid body, to whose center of figure or attraction all motions are to be referred, by physical considerations. The drift of his reasoning appears in the following extracts from his inaugural lecture On the Principles of the Galileo-

klidean space. It is probably not without significance that in the second edition of his Critique of Pure Reason Kant omits the third paragraph of the first section of the Transcendental Aesthetics, in which he had enforced the necessity of assuming the *a priori* character of the idea of space by the argument that without this assumption the propositions of geometry would cease to be true apodictically, and that "all that could be said of the dimensions of space would be that *thus far* no space had been found which had more than three dimensions."

* Naturgeschichte des Himmels, Werke, vol. vi, p. 152.

Newtonian Theory:* "The principles of the Galileo-Newtonian theories consist in two laws—the law of inertia proclaimed by Galileo, and the law of attraction added by Newton. . . . A material point, when once set in motion, free from the action of an extraneous force, and wholly left to itself, continues to move in a straight line so as to describe equal spaces in equal times. Such is Galileo's law of inertia. It is impossible that this proposition should stand in its present form as the corner-stone of a scientific edifice, as the starting-point of mathematical deductions. For it is perfectly unintelligible, inasmuch as we do not know what is meant by 'motion in a straight line,' or, rather, inasmuch as we do know that the words 'motion in a straight line' are susceptible of various interpretations. A motion, for instance, which is rectilinear as seen from the earth, would be curvilinear as seen from the sun, and would be represented by a different curve as often as we change our point of observation to Jupiter, to Saturn, or another celestial body. In short, every motion which is rectilinear with reference to one celestial body will appear curvilinear with reference to another celestial body. . . .

"The words of Galileo, according to which a material point left to itself proceeds in a straight line, appear to us, therefore, as words without meaning—as expressing a proposition which, to become intelligible, is in need of a definite background. *There must be given in the universe some special body as the basis of our comparison, as the object in reference to which all motions are to be estimated*; and only when such a body is given shall we be able to attach to those words a defi-

* Ueber die Principien der Galileo-Newton'schen Theorie. Leipzig, B. G. Teubner, 1870.

nite meaning. Now, what body is it which is to occupy this eminent position? Or, are there several such bodies? Are the motions near the earth to be referred to the terrestrial globe, perhaps, and those near the sun to the solar sphere? . . .

"Unfortunately, neither Galileo nor Newton gives us a definite answer to this question. But, if we carefully examine the theoretical structure which they erected, and which has since been continually enlarged, its foundations can no longer remain hidden. *We readily see that all actual or imaginable motions in the universe must be referred to one and the same body.* Where this body is, and what are the reasons for assigning to *it* this eminent, and, as it were, sovereign position, these are questions to which there is no answer.

"*It will be necessary, therefore, to establish the proposition, as the first principle of the Galileo-Newtonian theory, that in some unknown place of the universe there is an unknown body—a body absolutely rigid and unchangeable for all time in its figure and dimensions. I may be permitted to call this body* 'THE BODY ALPHA.' *It would then be necessary to add that the motion of a body would import, not its change of place in reference to the earth or sun, but its change of position in reference to the body Alpha.*

"From this point of view the law of Galileo is seen to have a definite meaning. This meaning presents itself as a second principle, which is, that a material point left to itself progresses in a straight line—proceeds, therefore, in a course which is rectilinear in reference to the body Alpha."

After thus showing, or attempting to show, that the reality of motion necessitates its reference to a rigid body unchangeable in its position in space, Neumann seeks to verify this assumption by asking himself the question, what consequences would ensue, on the hypothesis of the mere relativity of motion, if all bodies but one were annihilated. "Let us suppose," he says, "that among the stars there is one which consists of fluid matter, and which, like our earth, is in rotatory motion round an axis passing through its center. In consequence of this motion, by virtue of the centrifugal forces developed by it, this star will have the form of an ellipsoid. What form, now, I ask, will this star assume if suddenly all other celestial bodies are annihilated?

"These centrifugal forces depend solely upon the state of the star itself; they are wholly independent of the other celestial bodies. These forces, therefore, as well as the ellipsoidal form, will persist, irrespective of the continued existence or disappearance of the other bodies. But, if motion is defined as something relative —as a relative change of place of two points—the answer is very different. If, on this assumption, we suppose all other celestial bodies to be annihilated, nothing remains but the material points of which the star in question itself consists. But, then, these points do not change their relative positions, and are therefore at rest. It follows that the star must be at rest at the moment when the annihilation of the other bodies takes place, and therefore must assume the spherical form taken by all bodies in a state of rest. A contradiction so intolerable can be avoided only by abandoning the assumption of the relativity of motion, and conceiving motion as absolute, so that thus we are again led to the principle of the body Alpha."

Now, what answer can be made to this reasoning of Professor Neumann? None, if we grant the admissibility of the hypothesis of the annihilation of all bodies in space but one, and the admissibility of the further assumption that an absolutely rigid body with an absolutely fixed place in the universe is possible. But such a concession is forbidden by the universal principle of relativity. In the first place, the annihilation of all bodies but one would not only destroy the *motion* of this one remaining body and bring it to rest, as Professor Neumann sees, but would also destroy its very *existence* and bring it to naught, as he does not see. A body can not survive the system of relations in which alone it has its being; its *presence* or *position* in space is no more possible without reference to other bodies than its *change of position or presence* is possible without such reference. As has been abundantly shown, all properties of a body which constitute the elements of its distinguishable presence in space are in their nature relations and imply terms beyond the body itself.

In the second place the absolute fixity in space attributed to the body Alpha is impossible under the known conditions of reality. The fixity of a point in space involves the permanence of its distances from at least four other fixed points not in the same plane. But the fixity of these several points again depends on the constancy of their distances from other fixed points, and so on *ad infinitum*. In short, the fixity of position of any body in space is possible only on the supposition of the absolute finitude of the universe; and this leads to the theory of the essential curvature of space, and the other theories of modern transcendental geometry, which will be discussed hereafter.

There is but one issue from the perplexities of Euler, and that is through the proposition that the reality of rest and motion, far from presupposing that they are absolute, depends upon their relativity. The source of these perplexities is readily discovered. It is to be found in the old metaphysical doctrine, that the Real is not only distinct from, but the exact opposite of, the Phenomenal. Phenomenalities are the deliverances of sense; and these are said to be contradictory of each other, and therefore delusive. Now, the truth is that there is no physical reality which is not phenomenal. The only test of physieal reality is sensible experience. And the assertion, that the testimony of the senses is delusive, in the sense in which this assertion is made by the metaphysicians, is groundless. The testimony of the senses is conflicting only because the momentary deliverance of each sense is fragmentary and requires control and rectification, either by other deliverances of the same sense, or by the deliverances of the other senses. When the traveler in the desert sees before him a lake which continually recedes and finally disappears, proving to be the effect of *mirage*, it is said that he is deceived by his senses, inasmuch as the supposed body of water was a mere appearance without reality. But the senses were not deceptive. The lake was as real as the image. The deception lay in the erroneous inferences of the traveler, who did not take into account all the facts, forgetting (or being ignorant of) the refraction of the rays proceeding from the real object, whereby their direction and the apparent position of the object were changed. The true distinction between the Apparent and the Real is that the former is a partial deliverance of sense which is mistaken for the whole deliverance. The deception or illusion results from the circumstance that the senses are not properly and exhaustively interrogated and that their whole story is not heard.

The coercive power of the prevailing ontological notions of Euler's time over the clear intellect of the great mathematician is most strikingly exhibited in his statement that without the assumption of absolute space and motion there could be no laws of motion, so that all the phenomena of physical action would become uncertain and indeterminable. If this argument were well founded, the same consequence would follow, *a fortiori*, from his repeated admissions in the first chapter of his book, to the effect that we have no actual knowledge of rest and motion, except that derived from bodies at rest or in motion in reference to other bodies. Euler's proposition can have no other meaning than this, that the laws of motion can not be established or verified unless we know its absolute direction and its absolute rate. But such knowledge is by his own showing unattainable. It follows, therefore, that the establishment and verification of the laws of motion are impossible. And yet no one knew better than Euler himself that all experimental ascertainment and verification of dynamical laws, *like all acts of cognition*, depend upon the insulation of phenomena; that they can be effected only by disentangling the effects of certain forces from the effects of other forces (determinable *aliunde*, i. e., by their other effects) with which they are complicated—a proceeding which, in many cases, is facilitated by the circumstance that these latter effects are inappreciably small. Surely the verification of the law of inertia by the inhabitants of our planet does not depend upon their knowledge, at any moment, of the exact rate of its angular velocity of motion round the sun! And the validity of the Newtonian theory of celestial motion is not to be drawn in question because its author suggests that the center of gravity of our solar system moves in some elliptic orbit whose elements are not only unknown, but will probably never be discovered! As well might it be contended that the mathematical theorems respecting the properties of the ellipse are of doubtful validity, since no such curve is accurately described by any celestial body or can be exactly traced by a human hand!

Although in particular operations of thought we may be constrained, for the moment, to treat the Complex as simple, the Variable as constant, the Transitory as permanent, and thus in a sense to view phenomena "*sub quadam specie absoluti*," * nevertheless there is no truth in the old ontological maxim that the true nature of things can be discovered only by divesting them of their relations—that to be truly known they must be known as they are in themselves, in their absolute essence. Such knowledge is impossible, all cognition being founded upon a recognition of relations; and this impossibility nowhere stands out in stronger relief than in the exposition, by Newton and Euler, of the reality of rest and motion under the conditions of their determinability.

It follows, of course, from the essential relativity of rest and motion, that the old ontological disjunction between them falls, and that in a double sense rest differs from motion, in the language of Euler, "as one motion differs from another," † or, as modern mathematicians and physicists express it, that "rest is but a special case of motion." ‡ And it follows, furthermore,

* "De naturâ rationis est res sub quadam æternitatis specie percipere." Spinoza, Eth., Pars. ii, Prop. xliv, Coroll. 2.

† "Neque motus a quiete aliter differt, atque alius motus ab alio." Theoria motûs, etc., p. 8.

‡ "Die Ruhe ist nur ein besonderer Fall der Bewegung." Kirchhoff, Vorlesungen ueber math. Physik, p. 32.

that rest is not the logically or cosmologically *primum* of material existence—that it is not the natural and original state of the universe which requires no explanation while its motion, or that of its parts, is to be accounted for. What requires, and is susceptible of, explanation is always a change from a given state of relative rest or motion of a finite material system; and the explanation always consists in the exhibition of an equivalent change in another material system. The question respecting the origin of motion in the universe as a whole, therefore, admits of no answer, because it is a question without intelligible meaning.

The same considerations which evince the relativity of motion also attest the relativity of its conceptual elements, space and time. As to space, this is at once apparent. And of time, "the great independent variable" whose supposed constant flow is said to be the ultimate measure of all things, it is sufficient to observe that it is itself measured by the recurrence of certain relative positions of objects or points in space, and that the periods of this recurrence are variable, depending upon variable physical conditions. This is as true of the data of our modern time-keepers, the clock and chronometer, as of those of the clepsydra and hour-glass of the ancients, all of which are subject to variations of friction, temperature, changes in the intensity of gravitation, according to the latitude of the places of observation, and so on. And it is equally true of the records of the great celestial time-keepers, the sun and the stars. After we have reduced our apparent solar day to the mean solar day, and this, again, to the sidereal day, we find that the interval between any two transits of the equinoctial points is not constant, but becomes irregular in consequence of nutation, of the precession of the equinoxes, and of numerous other secular perturbations and variations due to the mutual attraction of the heavenly bodies. The constancy of the efflux of time, like that of the spatial positions which serve as the basis for our determination of the rates and amounts of physical motion, is purely conceptual.

The relativity of mass has repeatedly been adverted to in the preceding chapters. It has been shown that the measure of mass is the reciprocal of the amount of acceleration produced in a body by a given force, while force, in turn, is measured by the acceleration produced in a given mass. It is readily seen that the concept mass might be expanded, so as to assign the measure of mass, not to mechanical motion alone, but to physical action generally, including heat and chemical affinity. This would lead to an equivalence of masses differing with the nature of the agency selected as the basis of the comparison. Thermally equivalent masses would be the reciprocals of the specific heats of masses as now determined; and chemically equivalent masses would be the atomic weights, so called. It is important to note that the determination of masses on the basis of gravitation, in preference to their valuation on the basis of thermal, chemical or other physical action, is a mere matter of convenience, and is not in any proper sense founded on the nature of things.

But, apart from this, and looking to the ordinary method of determining the mass of a body by its weight, the relativity of mass is equally manifest. The weight of a body is a function, not of its own mass alone, but also of that of the body or bodies by which it is attracted, and of the distance between them. A body whose weight, as ascertained by the spring-balance or pendulum, is a pound on the surface of the earth, would weigh but two ounces on the moon, less than one fourth of an ounce on several of the smaller planets, about six ounces on Mars, two and one half pounds on Jupiter, and more than twenty-seven pounds on the sun. And while the fall of bodies, *in vacuo*, near the surface of the earth amounts to about sixteen feet (more or less, according to the latitude) during the first second, their corresponding fall near the surface of the sun is more than four hundred and thirty-five feet.

The thoughtlessness with which it is assumed by some of the most eminent physicists that matter is composed of particles which have an absolute primordial weight persisting in all positions and under all circumstances, is one of the most remarkable facts in the history of science. "The absolute weight of atoms," says Professor Redtenbacher,* "is unknown" — his meaning being, as is evident from the context, and from the whole tenor of his discussion, that our ignorance of this absolute weight is due solely to the practical impossibility of insulating an atom, and of contriving instruments delicate enough to weigh it.

There is nothing absolute or unconditioned in the world of objective reality. As there is no absolute standard of quality, so there is no absolute measure of duration, nor is there an absolute system of coördinates in space to which the positions of bodies and their changes can be referred. A physical *ens per se* and a physical constant are alike impossible, for all physical existence resolves itself into action and reaction, and action imports change.

* Dynamidensystem (Mannheim, Bassermann, 1857), p. 14.

CHAPTER XIII.

THE THEORY OF THE ABSOLUTE FINITUDE OF THE WORLD AND OF SPACE.—THE ASSUMPTION OF AN ABSOLUTE MAXIMUM OF MATERIAL EXISTENCE AS A NECESSARY COMPLEMENT TO THE ASSUMPTION OF THE ATOM AS ITS ABSOLUTE MINIMUM.—ONTOLOGY IN MATHEMATICS.—THE REIFICATION OF SPACE.—MODERN TRANSCENDENTAL GEOMETRY.—NON-HOMALOIDAL (SPHERICAL AND PSEUDO-SPHERICAL) SPACE.

It was shown in the last chapter, that the theory, according to which space and motion are real only on condition of being absolute, involved the assumption of the existence of an absolutely fixed point of reference in space, and that this again of necessity led to the doctrine of the absolute finitude of the universe. Although the connection between this doctrine and the prevalent ontological theorems respecting space and motion has not hitherto, so far as I am aware, been pointed out, the doctrine itself has been variously suggested in the interests of cosmological speculations founded upon the atomo-mechanical theory, as a means of escape from certain inevitable consequences of this theory with which those speculations are found at last to conflict. And it has recently been urged by eminent mathematicians upon considerations respecting the true nature of space and the real character of space-relations.

It is readily seen that the assertion of the absolute

finitude of the material universe is a logically integral part of the general assertion that whatever is real is absolute, and that the assumption of an absolute maximum of material existence is a necessary complement of the assumption of its absolute minimum, the atom. The first explicit announcement of a scientific belief in this maximum appears to have been made by C. F. Gauss, in one of his letters to Schumacher,* in which he discusses the attempts of his Transylvanian friend Bolyai and of the Russian geometer Lobatschewsky to found a geometrical system which should be independent of the Euclidean axioms in regard to parallels. The hints thrown out by Gauss in the letters just referred to, as well as in various parts of his other writings,† have, within the last twenty years, been fruitful of a discussion respecting the nature of space, the foundations of geometry and the origin and import of geometrical axioms, which has already produced an extensive and rapidly increasing literature.‡ The first effective

* Gauss, Briefwechsel mit Schumacher, vol. ii, pp. 268–271.

† Cf. "Disquisitiones generales circa seriem infinitam $1+\frac{\alpha\beta}{1\gamma}x+$," etc. (Comm. recent. Soc. Gott., ii, 1811–'13); "Theoria residuorum biquadraticorum Commentatio secunda" (*ib.*, vii, 1828–'32). To those who are familiar with Herbart's theory that our idea of spatial extension is a psychological elaborate of qualitative data, i. e., of sensations which are in themselves without extension, it will not appear improbable that Gauss's mathematical transcendentalism was to some extent due to the speculations of his colleague in the philosophical faculty of Goettingen, although Gauss habitually professed great contempt for the Herbartian system—just as Descartes was influenced by the teachings of his antagonist Gassendi. The connection of Gauss's metageometrical or (to use the expression of Lobatschewsky) pangeometrical views with his investigations respecting the geometrical interpretation of imaginary quantities and the theory of "complex numbers" is apparent.

‡ Cf. Halstead, Bibliography of Hyper-Space and non-Euclidean Geometry. American Journal of Mathematics, vol. i, pp. 261 *seq.* and 384 *seq.*; *ib.*, vol. ii, p. 65 *seq.*

impulse to this new departure in the walks of mathematical theory was given by Riemann in a remarkable dissertation * read before the philosophical faculty of Goettingen, June 10, 1854 (published by Dedekind in 1866, after Riemann's death), and by Helmholtz in an equally remarkable essay † published two years later. These publications have since been followed by numerous articles, pamphlets and books expository of the doctrines thus advanced, and, as was to be expected, there has been no lack of writings in which these doctrines have met with criticism and denial.

The articles of the new geometrical faith are certainly startling. Among them are propositions such as these: that our ordinary "Euclidean" tridimensional and "homaloidal" (flat) space is but one of several possible forms of space; that the preëminence of this Euclidean space among other forms of space can be maintained upon empirical grounds alone, and, in the sense of the logical and psychological tenets of the sensationalist school, depends solely upon the accidents of notional association, which may be (and, in the opinion of some enthusiastic advocates of the new doctrines, have been) overthrown by the discovery that the existence of additional dimensions is a necessary inference from certain facts of experience which can not otherwise be explained—just as the third dimension of space is said to be, not directly perceived, but simply inferred from familiar facts of visual or tactual experience for whose explanation the third dimension is an indispensable

* Ueber die Hypothesen welche der Geometrie zu Grunde liegen (Abhandlungen der Kgl. Gesellschaft der Wissenschaften zu Goettingen, vol. xiii, p. 133 *seq.*).

† Ueber die Thatsachen die der Geometrie zu Grunde liegen (Nachrichten der Kgl. Gesellschaft der Wissenschaften zu Goettingen, 1865, June 3).

postulate; that true and real space, therefore, has, or at least, for aught we know, may have, not three but four or even a greater number of dimensions; that the space in which we move is, or may be, not homaloidal or flat, but essentially non-homaloidal, curved, spherical or pseudo-spherical, so that every line, which we have hitherto regarded as straight, may upon sufficient prolongation prove to be a closed curve; that, by reason of the inherent and essential curvature of space, the universe, though unlimited, may be, and probably is, not infinite, but finite; that on the supposition of the pseudo-spherical character of space, a whole pencil of "shortest lines" may be drawn through the same point, all which are parallel to a given other "shortest line" in the sense that they will never intersect with it, however far produced; that not only the measure of curvature of space, as well as the number of its dimensions, may be, and probably is, different in different spatial regions, so that no valid inference can be drawn, from our experiences in the regions in which we happen to dwell, as to the curvature or the dimensions of space immeasurably distant or immeasurably small, but that in any given region both the curvature of space and the degree or number of its dimensions may be, and probably is, undergoing a gradual transformation, and so on.*

* The more cautious pangeometers have of late evinced a disposition to stigmatize some of the doctrines above enumerated, particularly those relating to the increase in the number of spatial dimensions and to the local differences and changes in the constitution of space, as inventions of their enemies or as extravagances of persons who are carried away by their enthusiasm. I may be pardoned, therefore, for citing a passage from a lecture of Professor P. G. Tait (who is certainly ready enough, as the book I quote from shows, to insist on sobriety in physics and mathematics at least, whatever may, in his opinion, be the appropriate frame

However dissonant from the teachings of our familiar experience these propositions seem, it is claimed that they are by no means without empirical warrant.

of mind for surveying the "Unseen Universe"): "The properties of space," says Tait, "involving (we know not why) the essential element of three dimensions, have recently been subjected to a careful scrutiny by mathematicians of the highest order, such as Riemann and Helmholtz; *and the result of their inquiries leaves it as yet undecided whether space may or may not have precisely the same properties throughout the universe.* To obtain an idea of what is meant by such a statement, consider that in crumpling a leaf of paper, which may be taken as representing space of two dimensions, we may have some portions of it plane, and other portions more or less cylindrically or conically curved. But an inhabitant of such a sheet, though living in space of two dimensions only, and therefore, we might say beforehand, incapable of appreciating the third dimension, would certainly feel some difference of sensations in passing from portions of his space which were less to other portions which were more curved. *So it is possible that, in the rapid march of the solar system through space, we may be gradually passing to regions in which space has not precisely the same properties as we find here—where it may have something in three dimensions analogous to curvature in two dimensions—something, in fact, which will necessarily imply a fourth-dimension change of form in portions of matter in order that they may adapt themselves to their new locality.*" P. G. Tait, On Some Recent Advances in Physical Science, p. 5. In keeping with this passage is a note of the distinguished mathematician, Professor J. J. Sylvester, to his opening address to the Mathematical and Physical Section of the British Association at Exeter, in 1869, as follows: "It is well known, to those who have gone into these views, that the laws of motion accepted as a fact suffice to prove in a general way that the space we live in is a flat or level space (a 'homaloid'), our existence therein being assimilable to the life of a bookworm in the flat space; *but what if the page should be undergoing a process of gradual bending into a curved form?* Mr. W. K. Clifford has indulged in some remarkable speculations as to the possibility of our being able to infer, from certain unexplained phenomena of light and magnetism, the fact of our level space of three dimensions being in the act of undergoing in space of four dimensions (space as inconceivable to us as our space to our supposititious bookworm) a distortion analogous to the rumpling of the page. I know there are many who, like my honored and deeply lamented friend, the late eminent Professor Donkin, regard the alleged notion of generalized space as only a disguised form of algebrai-

It is insisted that there are numerous optic, magnetic and other physical phenomena of which they yield the only sufficient explanation. Moreover, it is said that they alone afford a clew to the mysteries of modern spiritism, enabling us to bring within the chain of natural causation certain magical performances which we should otherwise be constrained to relegate to the regions of the Supernatural. In the first article of the first number of the American Journal of Mathematics, Professor Simon Newcomb demonstrates analytically that, "if a fourth dimension were added to space, a closed material surface (or shell) could be turned inside out by simple flexure without either stretching or tearing," Felix Klein having shown, some time before, that knots can not exist in a four-dimensional space. Accordingly, Professor Zoellner accounts for the well-known feats of the American "medium" Slade on the principle of the fourth dimension—one of these feats, however, strangely enough, consisting in the production of real trefoil knots in a rope the ends of which were sealed together and held in Zoellner's hands. And, finally, it is asserted that the theorems of Lobatschewsky, Riemann,

cal formulization; but the same might be said with equal truth of our notion of infinity in algebra, or of impossible lines, or lines making a zero angle in geometry, the utility of dealing with which as positive substantiated notions no one will be found to dispute. Dr. Salmon, in his extensions of Chasles's theory of characteristics to surfaces, Mr. Clifford in a question of probability, and myself in my theory of partitions, and also in my paper on Barycentric Projection, in the Philosophical Magazine, have all felt and given evidence of the practical utility of handling space of four dimensions as if it were conceivable space. Moreover, it should be borne in mind, that every perspective representation of figured space of four dimensions is a figure in real space, and that the properties of figures admit of being studied, to a great extent, if not completely, in their perspective representations." Nature, vol. i, p. 237 *seq.* The italics in the above passages are mine.

Helmholtz and Beltrami,* are the only true basis of a proper and exhaustive theory of parallelism. In the fullness of their faith in the impregnability of these positions, the votaries at the shrine of geometrical transcendentalism make bold to announce that, with the appearance of Lobatschewsky's "Geometrical Investigations," † a new era has dawned upon the mathematical world, and that in the daylight of this era the whole body of geometrical truths will be reduced to simplicity and order in a way analogous to that in which the theory of celestial motions was simplified and cleared up by the great thought of Copernicus. "What Vesalius was to Galen," exclaims Professor Clifford,‡ "what Copernicus was to Ptolemy, that was Lobatschewsky to Euclid."

The debate between the disciples of the new transcendental or pangeometrical school and the adherents of the old geometrical faith presents one feature which can not fail to strike the ordinary observer with some amazement. The disciples of the new school take their stand firmly upon empirical ground; their very first proposition is that all geometrical truths are of empirical origin, and that all we know of space and its properties is what we are taught by sensible experience. This proposition and the consequent denial of the

* An Italian mathematician who has investigated the properties of pseudo-spherical surfaces, which are distinguished from other surfaces of constant curvature by the fact that they admit of a sort of parallelism, in the transcendental sense, between their "straightest lines." A reference to Beltrami's writings and a brief exposition of their contents may be found in Helmholtz's essay on "The Origin and Meaning of Geometrical Axioms," Mind, vol i, p. 306.

† Geometrische Untersuchungen zur Theorie der Parallellinien, von Nicolaus Lobatschewsky. Berlin, Fincke'sche Buchhandlung, 1840.

‡ Philosophy of the Pure Sciences, W. K. Clifford's Lectures and Essays, vol. i, p. 297.

transcendental origin of geometrical axioms are emphasized by Riemann and Helmholtz alike. And yet, upon this foundation they construct a theory which lands us in the remotest regions of transcendentalism—in the realms of a metageometrical space in which all our wonted powers of imagination and conception are at fault and in which the facts of every-day experience as well as their natural relations are wholly out of sight. On the other hand, the most conspicuous champions of the old geometrical creed, in their defense of the familiar data of sensible experience and in their antagonism to the "vagaries" of transcendental geometry, invoke the doctrine of the non-empirical or transcendental origin of our ideas of space and its essential relations. The pangeometers erect a transcendental structure on empirical foundations, while the ordinary geometers build a system conforming to the data of experience upon transcendental grounds. This circumstance, however, strange as it appears at first sight, will hardly surprise the thoughtful student of the history of theories of cognition, or the intelligent reader of the preceding pages. It is by no means unusual to find that ontological speculations, whether they appear in the guise of physical or in that of metaphysical theories, prove subversive in the end, not merely of the facts for whose explanation they were devised, but of the very supports by which they are supposed to be upheld.

Having indicated, generally, the purport and scope of the transcendental theory of space, I now proceed to the examination of the premisses upon which it rests and of the arguments by which it is sought to be sustained. Here, at the outset, we find an assumption which obviously lies at the base of the whole theory: the assumption that space is a physically real thing—

not merely an object of experience, but an independent object of direct sensation whose properties may be ascertained by the aid of the ordinary instruments of physical and astronomical research—whose degree of curvature, for instance, is to be determined by means of the telescope. This assumption is explicitly stated by each of the three great expounders of the theory in question. "The only means at our command," says Lobatschewsky,* "to determine the accuracy of the propositions (calculations) of ordinary geometry consist in an appeal to astronomical observations." To the same effect Riemann: † "If we assume that bodies exist independently of their location in space, the measure of curvature (of space) is everywhere constant; *and then it follows from astronomical measurements* that it is not different from zero." And in the same sense Helmholtz: ‡ "All systems of practical mensuration that have been used for the angles of large rectilinear triangles, and especially all systems of astronomical measurement which make the parallax of immeasurably distant fixed stars equal to zero (in pseudo-spherical space the parallax even of infinitely distant points would be positive), confirm empirically the axiom of parallels and show the measure of curvature of our space thus far to be indistinguishable from zero. *It remains, however, a question, as Riemann observed, whether the result might not be different if we could use other than our limited base-lines, the greatest of which is the major axis of the earth's orbit.*"

The view thus taken of the nature of space and of

* Geometrische Untersuchungen, etc., p. 60.

† Ueber die Hypothesen, etc., Abhandl. der Kgl. Gesellschaft der Wissenschaften zu Goettingen, vol. xiii, p. 148.

‡ "On the Origin and Meaning of Geometrical Axioms," Mind, vol. i, p. 314.

the origin of our notions concerning it is obviously indicative of a decided advance beyond the farthest outposts of the old sensationalist camp. Nevertheless, it is supported in the main by a reference to the writings of a British thinker, J. S. Mill, who has been repeatedly referred to in these pages, and who is regarded, especially on the Continent, as the ablest modern expounder and defender of the doctrines of sensationalism, so far, at least, as they bear upon the subject now under consideration.* Stated in brief words, these doctrines are that the idea or notion of space is directly derived from sensible experience; that the properties of space are to be determined by observation or experiment; that the fundamental truths of geometry, like all other truths of physical science, are of inductive origin and warrant; and that the certainty to be attributed to geometrical theorems, though possibly different in degree, is not different in kind from that belonging to any general assertion respecting physical facts. The peculiar tenets of pangeometry being thus founded, in great part at least, upon the general sensationalist theory, it will be useful to enter upon a brief examination of this theory before I proceed to discuss the pangeometrical tenets themselves. For this pur-

* I do not mean to say that Riemann and Helmholtz themselves directly refer to Mill. But there are few German physicists and mathematicians who have not been diligent students of Mill's Logic, particularly since the appearance of Schiel's translation and the extravagant praises of Liebig; and this is quite apparent in most of the writings of the pangeometers. The interest with which each new edition of Mill's Logic has been received by scientific men everywhere is mainly due, doubtless, to its frequent references to scientific methods and results. The fact is that Mill has, for a series of years, been the official logician and metaphysician of the Continental naturalists and mathematicians, and the regard in which he is held by contemporary men of science is not unlike that which Aristotle enjoyed among the early mediæval scholastics.

pose I select an exposition of the theory in the book above referred to, the System of Logic of J. S. Mill, in which the fifth chapter of the second book "On Demonstration and Necessary Truth" contains an elaborate statement of the author's views on the basis and methods of geometrical science.

"The foundation of all sciences, even deductive or demonstrative sciences," says Mill* "is Induction; every step in the ratiocination of geometry is an act of induction. . . . The character of necessity ascribed to the truths of mathematics, and even (with some reservations to be hereafter made) the peculiar certainty attributed to them is an illusion; in order to sustain which it is necessary to suppose that those truths relate to, and express, the properties of purely imaginary objects. It is acknowledged that the conclusions of geometry are deduced, partly at least, from the so-called Definitions, and that those definitions are assumed to be correct representations, as far as they go, of the objects with which geometry is conversant. Now, we have pointed out that, from a definition as such, no proposition, unless it be one concerning the meaning of a word, can ever follow; and that what apparently follows from a definition follows in reality from an implied assumption that there exists a real thing conformable thereto. This assumption, in the case of the definitions of geometry, is not strictly true; there exist no real things exactly conformable to the definitions. There exist no points without magnitude; no lines without breadth, nor perfectly straight; no circles with all their radii exactly equal, nor squares with all their angles perfectly right. It will perhaps be said that the assumption does not extend to the actual, but only to the possible, exist-

* A System of Logic (eighth ed.), p. 168 *seq.*

ence of such things. I answer that, according to any test we have of possibility, they are not even possible. Their existence, so far as we can form any judgment, would seem to be inconsistent with the physical constitution of our planet at least, if not of the universe. To get rid of this difficulty, and at the same time to save the credit of the supposed system of necessary truth, it is customary to say that the points, lines, circles, and squares which are the subject of geometry exist in our conceptions merely, and are part of our minds; which minds, by working on their own materials, construct an *a priori* science, the evidence of which is purely mental, and has nothing whatever to do with outward experience. By however high authorities this doctrine may have been sanctioned, it appears to me psychologically incorrect. The points, lines, circles, and squares which any one has in his mind are (I apprehend) simply copies of the points, lines, circles, and squares which he has known in his experience. Our idea of a point I apprehend to be simply our idea of the *minimum visibile*, the smallest portion of surface which we can see. A line, as defined by geometers, is wholly inconceivable. We can reason about a line as if it had no breadth; because we have a power, when a perception is present to our senses, or a conception to our intellects, of *attending* to a part only of that perception or conception, instead of the whole. But we can not *conceive* a line without breadth; we can form no mental picture of such a line; all the lines which we have in our minds are lines possessing breadth. If any one doubts this, we may refer him to his own experience. I much question if any one, who fancies that he can conceive what is called a mathematical line, thinks so from the evidence of his consciousness: I suspect it is rather because he supposes that, unless such a conception were possible, mathematics could not exist as a science: a supposition which there will be no difficulty in showing to be entirely groundless.

"Since, then, neither in nature, nor in the human mind, do there exist any objects exactly corresponding to the definitions of geometry, while yet that science can not be supposed to be conversant about nonentities, nothing remains but to consider geometry as conversant with such lines, angles, and figures, as really exist; and the definitions, as they are called, must be regarded as some of our first and most obvious generalizations concerning those natural objects. The correctness of those generalizations, *as* generalizations, is without a flaw: the equality of all the radii of a circle is true of all circles, so far as it is true of any one: but it is not exactly true of any circle; it is only nearly true; so nearly that no error of any importance in practice will be incurred by feigning it to be exactly true. When we have occasion to extend these inductions, or their consequences, to cases in which the error would be appreciable—to lines of perceptible breadth or thickness, parallels which deviate sensibly from equidistance, and the like—we correct our conclusions by combining them with a fresh set of propositions relating to the aberration; just as we also take in propositions relating to the physical or chemical properties of the material if those properties happen to introduce any modification into the result; which they easily may, even with respect to figure and magnitude, as in the case, for instance, of expansion by heat. So long, however, as there exists no practical necessity for attending to any of the properties of the object except its geometrical properties, or any of the natural irregularities in those, it is convenient to neglect

S

the consideration of the other properties and of the irregularities, and to reason as if these did not exist: accordingly, we formally announce in the definitions that we intend to proceed on this plan. But it is an error to suppose, because we resolve to confine our attention to a certain number of the properties of an object, that we therefore conceive, or have an idea of, the object, denuded of its other properties. We are thinking, all the time, of precisely such objects as we have seen and touched, and with all the properties which naturally belong to them; but, for scientific convenience, we feign them to be divested of all properties, except those which are material to our purpose, and in regard to which we design to consider them.

"The peculiar accuracy, supposed to be characteristic of the first principles of geometry, thus appears to be fictitious. The assertions on which the reasonings of the science are founded do not, any more than in other sciences, exactly correspond with the fact; but we suppose that they do so, for the sake of tracing the consequences which follow from the supposition. The opinion of Dugald Stewart respecting the foundations of geometry is, I conceive, substantially correct; that it is built on hypotheses; that it owes to this alone the peculiar certainty supposed to distinguish it; and that in any science whatever, by reasoning from a set of hypotheses, we may obtain a body of conclusions as certain as those of geometry, that is, as strictly in accordance with the hypotheses, and as irresistibly compelling assent, *on condition* that those hypotheses are true."

I have quoted this passage, from Mill's Logic, at length, not only because it is the most elaborate and connected statement of the sensationalist theories concerning the character of necessary truths and especially the truths of geometry, but also because this statement exhibits certain peculiarities that are worthy of attention. One of these peculiarities is the concession that the mind has the power of abstraction, and of forming and reasoning about generalizations which, "*as* generalizations, are without a flaw." The inconsistency of this admission with the claim that "the points, lines, circles, and squares which any one has in his mind are simply copies of the points, lines, circles, and squares which he has known in his experience," is evident. And this inconsistency has not escaped the notice of other promoters of the experiential or sensationalist doctrine, as is shown, for instance, in the writings of Mr. Buckle, who does not hesitate to draw the true conclusions (from which Mill himself appears to recoil) from Mill's premisses. Buckle not only boldly asserts that there are no lines without breadth (he strangely forgets the thickness), but also that the neglect of this breadth by the geometrician vitiates all the results of geometrical inference, the only comfort vouchsafed to us being that the error, after all, is not very considerable. "Since, however," he says,* "the breadth of the faintest line is so slight as to be incapable of measurement, except by an instrument used under the microscope, it follows that the assumption that there can be lines without breadth is so nearly true that our senses, when unassisted by art, can not detect the error. Formerly, and until the invention of the micrometer, in the seventeenth century, it was impossible to detect it at all. Hence, the conclusions of the geometrician approximate so closely to truth that we are justified in accepting them as true. The flaw is too minute to be perceived.

* History of Civilization in England, vol. ii, p. 342 (Appletons' American edition).

But that there is a flaw appears to me certain. It appears certain that, whenever something is kept back in the premisses, something must be wanting in the conclusion. In all such cases, the field of inquiry has not been entirely covered; and, part of the preliminary facts being suppressed, it must, I think, be admitted that complete truth be unattainable, and that no problem in geometry has been exhaustively solved."

Whether Buckle was able to think of a line as the limit between two surfaces, and whether, in his opinion, such a limit has breadth (i. e., is itself a surface, so that we are driven from limit to limit, *ad infinitum*), he does not tell us. Nor does he say whether or not, in view of the fact that the breadth of a line depends upon the material out of which it is constructed, or upon which it is drawn, there ought to be a pasteboard geometry, a wooden geometry, a stone geometry, and so on, as distinct sciences.

To do justice, however, to Mill and the subject under discussion, we must keep before us Mill's own statement. Returning, then, to his exposition, the question arises at once: What does he mean when he says that none of the elements of space exist in fact as they are considered in the science of geometry—that, for example, there exist no lines perfectly straight? The only possible meaning is that none of the straight lines, so called, of which we have *experiential* knowledge, are congruent with the straight lines of which we have *other* knowledge—that they do not conform to the standard straight line in our minds. But Mill asserts that "the lines, etc., which any one has in his mind, are simply copies of the lines which he has known in his experience." There is no standard, therefore, with which the lines presented in experience can be compared and from which they can be shown to be divergent. Thus Mill's theory breaks down with the very first fact which he brings to its support.* This is no mere captious criticism. It is a simple exhibition of the utter senselessness of the premisses from which Mill's conclusions are drawn. The whole foundation of his theory crumbles the moment it is touched. And, upon further examination, it is found that he entirely mistakes the significance of the facts which he adduces. The real import of Mill's assertion just referred to is very different from that which he ascribes to it. The truth which lies at the bottom of that assertion is that we have no experiential knowledge, in Mill's sense, of lines, circles, squares, at all. We have experiential knowledge of so-called straight rods, cords, edges, or grooves, of spherical and cubical bodies with circular or square sections or sides; but our knowledge of points, lines, surfaces, and geometrical solids is obtained solely by the process of abstraction. Nothing is clearer and more readily demonstrable than that the elements of geometrical science—the foundations upon which the science of geometry rests—can not have been obtained by induction, and that, *a fortiori*, it can not be true, as Mill contends, that "every step in the ratiocinations of geometry is an act of induction." Induction is a cu-

* That so acute a thinker as J. S. Mill was blind to the many inconsistencies and absurdities with which his Logic and parts of his other writings abound, is explicable solely by the fact that he took his theory of cognition upon trust as a sacred inheritance from his father, who, in turn, had derived it from the French and English nominalists and sensationalists of the seventeenth and eighteenth centuries. The doctrines of these sensationalists were necessarily crude, because they originated at a time when rational psychology was in its infancy and comparative psychology was not even thought of; and they were extravagant because they were generated by the antagonisms of an equally extravagant realism.

mulation of instances in all of which the same element or feature is found along with other elements or features. But no one has ever seen two bodies whose edges, though called straight, did not prove to be broken in different degrees when examined with sufficient magnifying power. Experience furnishes no two instances presenting the feature of straightness in the same degree. Much less has any one seen a great number of bodies whose edges were exactly coincident. The same thing is, of course, true, *mutatis mutandis*, of points, curves, surfaces, and solids. The divergences between their figures as well as their magnitudes become more apparent in proportion to the magnifying power with which they are examined. And their true figures are wholly undiscoverable by any magnifying process at our command. The truth is, that we never get sight or come into the actual presence of a true and complete geometrical fact. It is simply nonsense, therefore, to say, with Mill, that the points, lines, surfaces, solids, etc., with which the science of geometry deals and respecting which it is able to draw valid deductions, are *real* (i. e., physical) and not imaginary points, lines, surfaces, and solids, and that the points, lines, surfaces, and solids in our minds are copies of them. It is true enough that the geometrical elements are not imaginary, because they have reference to real facts; nor are they in any proper sense *hypothetical*, as is contended by Dugald Stewart; they are *conceptual*, the results of abstraction. If it were otherwise, deductive geometrical ratiocination—and, indeed, any other kind of reasoning properly so called—would be utterly impossible. All deductive reasoning depends upon the power of abstraction. And this truth is applicable, not only to geometry and to mathematics generally, but to all sciences whatever. This is so for two reasons: In the first place, no physical thing (or historical event) ever becomes experientially known to us with all its properties, relations, or incidents; sensation and perception never furnish the intellect with a complete fact. And, in the second place, as I have heretofore shown, in dealing with the facts, so called, obtained at the hands of sensible experience, the intellect is restricted to certain definite relations which it segregates or abstracts from other relations. In the processes of discursive thought, the intellect never has before it either sensible objects or the whole complement of relations which make up their mental images or representations, but only some single relation or class of relations. It operates along *lines* of abstraction, the final synthesis of whose results never yields anything more than outlines of the objects represented. During all its operations the intellect is fully aware that neither any one link in the chain of abstraction nor the group of abstract results which we call a concept (in the narrower sense of a collection of attributes representing an object of intuition or sensation) is a copy or an exact counterpart of the object represented. It is always conscious that, to bring about true conformity between concepts or any of their constituent attributes with forms of objective reality, the group of relations embodied in these concepts would have to be supplemented with an indeterminate number of other relations which have not been apprehended and possibly are insusceptible of apprehension. But this nowise affects the validity of the intellectual operation. The mathematician, when he determines the properties of a conic section, knows full well that he will never meet with a body whose geometrical outline is an exact exemplification of the law of the constancy of the ratio between the distances of any one of its points from a fixed point and a fixed straight line respectively, and that there is in nature no trajectory which strictly coincides with such a curve. But this knowledge does not in the least degree disturb his faith in the perfect validity of his reasoning. When he comes to apply the results of this reasoning to a natural fact, he supplements it, as well as he can, with the results of other processes of reasoning based upon other known relations of the same fact, and thus approaches the fact as nearly as possible, nothing daunted by the ever-present reflection that he will never succeed in coming into the actual presence of the whole fact with all its relations.

It is obvious that the conformity of the results of abstract or conceptual reasoning to the data of experience is in direct proportion to the degree of independence of the relations dealt with from the other relations which constitute the conditions of real existence in the object represented in the operation of thought. Herein lies the preëminence of geometry among the physical sciences. In the physical sciences usually so called the relations about which these sciences are conversant are closely interdependent; the thermal, electric, magnetic, optical, and chemical properties of a body in various ways determine each other. If the nature and degree of this interdependence were accurately known and could be brought within the reach of exhaustive conceptual analysis, these sciences would become deductive to the same extent to which the science of geometry is deductive. All the physical sciences are constantly striving to progress in this direction, but the progress is so slow as to afford little hope that the goal here indicated will ever be reached. One reason for this is that the number of newly discovered relations multiplies at the same (if not at a greater) rate at which the nature and degree of the interdependence between the relations already known are brought to light. And the difficulty of determining the interdependence in question increases geometrically as the number of new relations is augmented arithmetically.

The foregoing reflections are sufficient, in my judgment, to show that the sensationalist views of space and of the nature and warrant of geometrical truth are untenable, at least in the form in which they are propounded by Mr. Mill. But these reflections do not in the least degree impugn the general proposition that all our knowledge of the objective world is derived from experience. This proposition appears to me to be undeniable, and is, doubtless, assented to, explicitly or in some mode of implication, by every sane person at the present day, the only controversies respecting it being disputes about the meaning of terms. But the sensationalists, and especially, as I have already shown, the founders and supporters of transcendental geometry, advance a thesis which is to be carefully distinguished from the proposition just stated. They maintain that space is not only objectively real, but a direct and independent object of sensation whose properties are to be empirically ascertained like those of any other physical thing. This assertion has been met with the counter-assertion, made by the antagonists of geometrical transcendentalism, that space, like time, is not an independent object of sensation, but, as Kant has taught, or is supposed to have taught, a mere form of intuition, a state or condition of the intellect existing independently and in advance of all sensible experience. The contest between the champions of the new doctrine and their opponents has been mainly carried on under the belief, common to both the contending parties, that these views are strict alternatives, and that no other view is admissible or possible. Let us test these two conflicting assertions by facts of cognition about which there is no contest, or which clearly can not be contested on rational grounds.

First, as to the assertion of Riemann and Helmholtz: if space is a physically real object, it certainly is not a thing outside of, coördinate with, and different from, other physical objects. When we say that all things are in space, we do not mean that they are contained in it as water is contained in a vessel, but we mean that there is no objectively real thing which is not spatially extended, or, according to the usual form of expression, that spatial extension is a primary property of all varieties of objective existence. This fact is so obvious that Descartes was led by it to maintain that spatial extension was the only true essence of objective reality. In what way, then, and by what means, do we distinguish between space and physical things ordinarily so called? Certainly not, or at least not directly, by sensation. Different acts of sensation may present different properties of the same object, and these properties may thus be dissociated. But no act of sensation dissociates the extension of a body from all its other properties and presents the property of extension alone. The sensationalists, however, contend (and here they trench upon the ground of their opponents, the Kantian idealists) that, although there are no physical objects without spatial extension, and although such extension is in a sense a common property of all physical objects, nevertheless these objects do not fill all space, there being pure space between them. The reply to this is that this assertion, if true, does not help the sensationalists. For acts of sensation are possible only when and where there is objective difference and change; we have direct sensation of different and variable physical qualities ordinarily so called, and not of that which is absolutely homogeneous and invariable. Here comes in Hobbes's law: *Sentire semper idem et non sentire ad idem recidunt.* It is precisely the fact of its homogeneity and unchangeableness, in addition to that of its invariable presence in all physical objects, which distinguishes the property of spatial extension from all other properties characteristic of a real thing, and enables the sensationalist to speak of the existence of space at all. If this distinction could be obliterated —if the cognitive or conceptual barrier which separates the sensations produced by physical action from the states of consciousness representative of space were once broken down—there would no longer be any ground whatever for the distinction between the "properties" of space and the properties of matter in any of its varieties. We should be constrained to say that the only form or variety of objective existence is either space or matter (it being a mere question of nomenclature which), and that all the properties we now attribute to matter are in truth and in fact properties of space.

That all this should have escaped the attention of Riemann and Helmholtz is marvelous, considering the assumption made by both in order to account for the alleged necessity of attributing to space a constant measure of curvature, and thus limiting the number of the species of space, which, according to their statement, are admissible, to three, viz.: spherical space with a positive curvature, pseudo-spherical space with a negative curvature, and flat or homaloidal space with a curvature equal to zero.* I allude to the assumption that bodies, in the language of Riemann already quoted, "exist independently of their location in space," which means, of course, that they are at least different from, if they have not a physical constitution wholly independent of, space. But for this assumption there can be no valid reason founded upon or consistent with the premisses of the transcendental theory, why space may not be essentially paraboloidal, or hyperboloidal, or polyhedral, or of any other inherent form evolvable from the creative fancy of the next non-homaloidal intellect.

This brings me to the allegation of the transcendentalists, that the properties of space, such as the degree and form of its curvature, are to be determined by experiment. How would such a determination be effected? Suppose an astronomer, at proper intervals, directed his telescope to some fixed star whose distance from the earth he knew in some way (say from spectroscopic data) to be far greater than that of *Arcturus*, for the purpose of ascertaining its parallax. And suppose he found this parallax sensibly to exceed that of the less distant star—in other words, suppose he found that the angle of intersection between the lines of his vision was different from that required by the known facts and laws of astronomy and optics: what would be his conclusion? It is not difficult to anticipate the answer to this question, for the case supposed is not without precedent in the history of astronomy. Displacements in the lines of vision have repeatedly been observed by astronomers, who were unable to account

* Felix Klein "(Ueber die Nicht-Euklidische Geometrie," Mathematische Annalen, vol. iv, p. 577) designates these kinds of space as *elliptic*, *parabolic*, and *hyperbolic*.

for them by the facts and natural laws of which they had knowledge. In the early part of the last century, Bradley (with the aid of Molyneux) made a series of telescopic observations of the star γ *Draconis*, to the end of determining the amount of its apparent displacement due to the orbital motion of the earth, so as to detect the annual parallax of the fixed stars—an achievement very desirable in Bradley's time by reason of a standing objection to the Copernican system urged on the ground of the alleged absence of such a parallax. To his surprise he found a displacement different in direction and far greater in degree than that looked for. This anomaly had to be explained; and Bradley knew of no physical cause to which it could be assigned. He thought for a time of nutation; then of refraction; but he soon became satisfied that no explanation was afforded by either fact. He was finally led, by a careful study of the variations in the direction and rate of the displacement, to look for a solution of the mystery in the composition between the velocity of light and that of the earth's orbital motion, and thus became the discoverer of what is now known as the aberration of light. Amid all his perplexities, however, it does not appear at any time to have occurred to him that the anomalous phenomenon could be the result of a constitutional crook in space. And it may be asserted with confidence that there is no astronomer living to-day who would attribute the anomalous parallax, whose discovery I have supposed, to a spatial pseudo sphericity. For, irrespective of all other considerations, the astronomer would at once meet every suggestion of this sort with the objection that an inherent curvature of space presupposes differences between its several parts—heterogeneities in its internal constitution—and that the hypothesis suggested, therefore, involved nothing less than the attribution to space of the very properties by the absence of which alone it is distinguishable from matter.

The theory of the geometrical transcendentalists is thus invalidated by the absurdity of its fundamental assumption. Space is not, can not be, an object of sensation. The attribution to space of relations and sensible interactions of the kind reflected in sensation is impossible without the assumption of diversities among its constituent parts, the denial of which is the basis of *every* notion or concept of space, whatever may be the logical or psychological doctrine to which that notion is referred. Are we driven, then, to the counter-assumption of the Kantian idealists, that space is a purely subjective form of intuition existing in the mind independently of and antecedently to all acts of sensation—to the doctrine of the metaphysical and mathematical adversaries of geometrical transcendentalism? Let us see by what arguments this doctrine is enforced.

The Kantian idealists affirm that the idea of space is not only an invariable element of every act of sensation, but a condition precedent to sensation; that, before we are able to refer any subjective impression to an objective cause, and thus to speak of the existence of objectively real things or phenomena at all, the basis of this reference—of the relation, not merely between the Within and the Without, but also between two elements at least of the Without whose interaction produces the sensation—must already be present in the intellect. Sensation, it is said, is of objects; it is essentially a step from a subjective affection or feeling to an objective reality. Where is the ground for this step? Not, contends the

Kantian, in the world of objects; for the objects are reached and become existent in intuition and sensation only by means of the step. It must be in the subject, therefore, in the intellect; and it must be present there in advance of the act of sensation. That this is so, appears, moreover (it is claimed), from the fact that the idea of space is absolutely indestructible. We can mentally evacuate space of its sensible contents; the intellect can "think away" everything that is an object of sensation; but it can not "think away" space itself. Space is an integral part of all states of consciousness whatever.

The foregoing exposition is a fair and sufficiently exhaustive statement of the Kantian view. This view has one feature in common with that of the sensationalists to which I have already made incidental allusion —the assumption that space exists, either as an object of sensation or as a form of intuition, *as an independent fact*, and is therefore susceptible of objective or subjective apprehension *by itself*. I have already shown that this assumption, in the sensationalist sense, is unfounded. And, upon careful examination, it proves equally unfounded in the sense of the idealists. It is not true that we can mentally evacuate space of all its contents, and have in the mind, or before the mind, the form or image of pure space. On the contrary, the idea of space is invariably associated in consciousness with some definite quality of sense. When we attempt to bring space before the mind (or, as it is usually called, to "realize" it) in its visual aspect, it always appears in synthesis with a mental reproduction of some sensation of color, however faint. Similarly, when we make the effort of mentally "realizing" or representing it in its tactual aspect, it proves equally indissociable from a reproduction of some form of pressure or feeling.* In this respect the arguments of Hume and Berkeley (which are of necessity simple appeals to consciousness) have never been successfully met. The dissociation between the "idea" of spatial extension and the feeling or feelings constituting sensation which we are able—and, for the purposes of discursive reasoning, constrained—to effect, is not an *intuitional*, but a *conceptual* dissociation. Whenever we contemplate and reason about an objectively real thing, we can, in virtue of the power of abstraction, attend to the property of spatial extension, in total disregard of its sensible qualities; but whenever we strive to bring its extension before the mind as real—to frame a mental image of extension, or to represent it as a distinct form of intuition—we are instantly forced to invest or associate it with some one datum of sensation which we interpret as the incident or reaction of a physical process. Intuition (using the word in the Kantian sense) is an integral part of sensation, and appears as such alike in the presentations of sense and their representations or reproductions in the phantasy.

This disposes of the Kantian argument that space must be a subjective form of intuition because the mind can not banish it from consciousness. And another simple reflection is equally fatal to the claim that space must be a subjective form *existing in advance of all acts of sensation*, inasmuch as it is the indispensable ground for the step by which the intellect reaches an object external to itself. The obvious answer to this is that, if space is purely subjective, being wholly *in* the mind, it certainly can not afford ground for a step *out*

* Cf. Sir William Hamilton's Lectures on Metaphysics, Lect. xxii; Stumpf, Ueber den psychologischen Ursprung der Raumvorstellungen (Leipzig, Hirzel, 1873), p. 19.

of it. This reflection is the true basis of the post-Kantian species of idealism, such as that of Fichte, and, in a sense, of Schopenhauer. But the whole argument, as well as the idealistic perplexities that have been occasioned by it, is founded on the old ontological assumption that things or entities exist independently of each other and otherwise than as terms of relations. That this is not true of objectively real things has been sufficiently enforced in the preceding pages of this book; and it is equally untrue of the relation between the cognizing subject and its object. In every act of primary cognition, the objective phenomenon, so called, and its subjective counterpart are born into consciousness at the same moment, because the reality of either depends upon that of the other. This is the great primary and irreducible fact of cognition, which is not the less a fact because it has been misinterpreted by the metaphysicians in a variety of ways, and has given rise to a host of absurd cognitive theories.

What, then, is the real nature of space and what is the true source of our knowledge respecting it? If the preceding considerations are valid and conclusive, this question admits of but one answer. Space is a concept, a product of abstraction. All objects of our sensible experience present the feature of extension in conjunction with a number of different and variable qualities attested by sensation; and, when we have successively abstracted these various sensations, we finally arrive at the abstract or concept of a form of spatial extension. I purposely say *form of extension*, and not simply *extension* or *space*, for the former, and not the latter, is the *summum genus* of the line of abstraction here indicated. If the word "concept" be used in the sense in which it is representative of a possible object of intuition, a *spa-*

tially extended form is the last result of the process by which an object or phenomenon can be conceived. The abstract or concept (using the word now in its wider sense) *extension generally*, or *space*, is reached by another series of abstractions of which I may have something to say hereafter. The failure to discriminate between those concepts, so called, which involve no reference to limits or forms and the true *summa genera* of the classification of sensible objects is one of the sources of the confusion which everywhere besets the theory of transcendental space, as we shall presently see.

The doctrines of the idealists (more properly called intellectualists) respecting the nature of space are, therefore, as untenable as those of the sensationalists. And the opinion of the disciples of Kant and Schopenhauer, that the teachings of transcendental geometry can be refuted by an appeal to the "Transcendental Æsthetics" of the "Critique of Pure Reason" is a mistake. The proposition that space is a pure subjective form of intuition, if true, could not in the slightest degree shake the position of the geometrical transcendentalist. His simple retort upon the Kantian is, that, if space is an innate form or condition of the intellect determining the apprehension of external objects in a certain order, or according to certain laws, it is again a question of fact, what is that order and what are those laws. Whether space be within the mind or without it, the question of its flatness, sphericity or pseudo-sphericity remains. Whether the form of the lines and surfaces possible in space is the result of its physical constitution outside of the mind, or of the internal constitution of the mind itself—in either case the fact is the same, whatever it may prove to be. This is in entire accord with Kant's own distinct declaration in his "Notes to Transcendental Æsthetics," * when he says that our mode of intuition is not *necessarily* confined to the peculiar constitution of our minds, but *may* be shared by other thinking beings, "although this is a matter which we are unable to decide." From this declaration the inference is unavoidable that the question, what the precise form of intuition is in a given intellect, is purely a question of fact. In this respect, then, Helmholtz † is unquestionably in the right as against Land, Krause, Becker, and the other Kantians.

Having reached the conclusion that space is neither a physical object of sensation, nor an innate form of the mind independent of and preëxisting to all sensation, but a concept, we are now able to enter upon a series of considerations akin to those already presented against the alleged experimental determinability of the curvature of space, by which the true character of the transcendental theory of space may be so thoroughly exhibited that there can be no rational controversy respecting its merits. The first of these considerations is this: If the doctrines of the transcendentalists are founded in fact, it follows that there is in space a coercive power resulting from its constitution which makes lines and surfaces other than those conforming to its inherent figure impossible. If space is not "flat," but spherical, for instance—I assume for the moment, and for the sake of argument, that there is sense in the assertion of the "flatness" of ordinary "Euclidean" space—then every line in it necessarily follows a definite course to which it is astricted by the internal law governing the

* Kritik der reinen Vernunft (ed. Rosenkranz), p. 49.

† Cf. "The Origin and Meaning of Geometrical Axioms," Mind, vol. iii, p. 212 *seq.*, also Die Thatsachen in der Wahrnehmung, Berlin, 1879.

arrangement of its parts. From this it is a legitimate and inevitable consequence that, in a space of definite and inherent curvature, lines even of different degrees of curvature are impossible. The measure of curvature of such a space being once determined, all its lines must conform to it. To this it is no answer to say that Lobatschewsky and Beltrami have shown the practicability of constructing consistent and logically coherent systems of geometry on the principle of the non-parallelism of "shortest lines," and that Professor Lipschitz has demonstrated that the laws of motion as dependent on motive forces could also be consistently transferred to spherical or pseudo-spherical space, so that the comprehensive expression for all the laws of dynamics, Hamilton's principle, may be directly transferred to spaces of which the measure of curvature is other than zero. For the constructions of Lobatschewsky and Beltrami (which serve also as the basis of Lipschitz's investigations) are all constructions of *lines* and *surfaces;* and these constructions are founded upon postulates utterly inconsistent with the properties of non-Euclidean space. One of these postulates is, that in spherical as well as in pseudo-spherical space it is possible to trace lines of any degree of curvature, and therefore also lines whereof the curvature is zero, that is to say, straight lines in the old sense. How, indeed, could the "measure of curvature" be otherwise determined? That measure depends upon the *radius* of curvature; according to Gauss, the measure of curvature belonging to every surface that admits of the motion of the figures lying upon it, without change of any of their lines and angles, measured along it, is the constant reciprocal of the greatest and least radii of curvature. These radii are *straight* radii, in the old sense; for, if they are not straight, they are of some definite degree of curvature, which again can be determined only by reference to another particular radius, and so on, either *ad infinitum*, or until we come at last to the old Euclidean straight line.

The legitimate premisses of the theory of non-Euclidean space lead to the inevitable conclusion that the lines of such a space, though curves, have neither tangents nor normals, neither radii nor cords, and that on the grounds of non-Euclidean postulates alone they are wholly indeterminable. This is again a curious exemplification of the ontological error according to which things and forms are determinable in themselves, without reference to, or contrast with, correlative things and forms. What is especially remarkable, in this aspect of the doctrine of the transcendentalists, is the ascription to real space of an inherent disjunction between the forms of its alleged curvature—the assertion that its measure of curvature must be *either* positive, *or* negative, *or* zero. This assertion is all the more remarkable by reason of the transcendentalist claim that the new doctrine has emancipated the old system of geometry from its arbitrary limitations, and is a widening, a logical expansion, of the idea of space.

The source of all the perplexities in which we find ourselves involved by the assumptions and theories of the transcendentalists is so obvious, that it is a wonder how it has come to be completely ignored by the adversaries of the new doctrine no less than by its adherents. The parent error of this doctrine is the assertion that the space, with which ordinary "Euclidean" geometry deals, is a "flat," and not a spherical or pseudo-spherical space. *The truth is that the space whose idea or notion underlies all geometrical constructions whatever, in-*

cluding those of the pangeometers, is neither flat, nor spherical, nor pseudo-spherical, nor of any other inherent figure, but is simply the intuitional and conceptual possibility of tracing any or all of the lines characteristic of plane, spherical, ellipsoidal, paraboloidal, hyperboloidal, etc., and, to some extent, pseudo-spherical surfaces within it—a possibility due to the circumstance that it is nothing more nor less than a concept formed by dismissing from our mental representation of physical objects, not only all the attributes constituting their physical properties other than extension, *but also all the determinations of figure by which they are distinguished.* This is the only sense in which we have any right to speak of space as even or homaloidal. Space has no internal structure or inherent figure, because it is not a physical object, and therefore has no "properties" which can be ascertained by experiment or observation. Nor has it any properties, rightly so called, that are determinable *a priori*, by an act of intuition. Space is one of those ultimates of abstraction in which the connotation coincides with the denotation, and in which, therefore, true connotation is at an end. I repeat: space has no properties, for, considered as an entity, it has no relations, its very essence being a denial of, or abstraction from, all relations. For this reason it is an abuse of terms to define geometry (as is so frequently done, and has lately been done by Professor Henrici *) as "the science whose object it is to investigate the properties of space." The object of geometry is the investigation of the possible determinations or limitations of space, i. e., of the relations between the various forms of extension or of the properties of figures.†

* Encycl. Britan., s. v. Geometry.

† In this sense D'Alembert (Élémens de Philosophie, § 15—Œuvres,

The whole science of geometry is conversant about that which the concept space of necessity excludes, viz., about determinations or limits. Geometry, indeed, has reference to space, inasmuch as the determinations with which it deals are spatial determinations. Upon this fact arises the difference between the scope of geometry and that of the other branches of pure mathematics, and the inapplicability of many of the methods and results of mathematical analysis to the relations between the forms of space—a difference the disregard of which is prolific of so many errors in the reasoning of those who seek to draw conclusions respecting the "properties" of space (such as the possible number of its dimensions) from the abstract concept *quantity*. Geometry is undoubtedly an empirical science, though not in the sense in which the term "empirical" is generally understood, and especially not in the sense in which it is interpreted by Mill and the geometrical transcendentalists. It is an empirical science, inasmuch as it deals with a property of physical things, *extension*, which is an ultimate, or, rather, primary and irreducible datum of the act of sensation—just as much such a datum as the sensation of color with which, as I have shown, the visual intuition of space is invariably associated. All attempts, such as those of Herbart, to produce the "idea" of extension by an elaboration of such data of sensation as are commonly designated as qualitative, are as abortive as the corresponding attempts to deduce the qualitative elements of sensation from forms of extension. The primary datum of extension is the empirical element in the science of geometry. This primary datum is not *space*, but *limited extension*, for sensation and intuition are of

tome i, p. 268) defines geometry as "la science des propriétés de l'étendue *en tant qu'on la considère comme simplement étendue et figurée.*"

particular bodies, and therefore of limited extension, not of extension generally, or space. Forms of limited extension, however, give rise to the concept space, by the application of the processes of abstraction I have indicated. On the other hand, the conclusions of geometry are not derived from empirical data alone, and are not reached by processes of induction, as Mill contends, and in that sense geometry is not an empirical science. *Nor is there any geometrical axiom which is purely a datum of sensation, as is asserted by the sensationalists, or of intuition, according to the teachings of the idealists or intellectualists.* All the geometrical axioms, which serve as starting-points of deduction, contain two elements: an element of intuition (as a part of sensation) and an element of arbitrary intellectual determination which is called *definition.* The facts of extension and its limits—surfaces, lines, and points—are given in intuition; without sensible experience we should not know anything about geometrical solids, surfaces, lines, and points; but nothing is deducible from the existence of these elements, or our intuition of them, until they are defined. This is evident upon a simple inspection of the geometrical axioms. The axiom that between two points but one straight line can be drawn (or, what is the same thing, that two straight lines can not inclose a space) involves the definition of the straight line—a definition, by the way, far more difficult on purely geometrical grounds than that of parallels.* Again: the axiom respecting parallels, in the

* The real source of this difficulty lies in a fundamental defect of the current theories of cognition—in the failure to see that all processes of deductive reasoning involve an ultimate reference to primary constants which are not given in experience, but established by the intellect. This primary constant in geometry is the straight line, or simple direction. That the difficulties presented by the 10th Euclidean axiom ("two straight

form now generally given to it, viz., that through a given point but one straight line can be drawn parallel to another straight line, presupposes the definition, not only of the straight line, but of parallelism which, in elementary geometry, presents the difficulty of involving the concept of infinite extension, and has given rise to innumerable quandaries (such as that of infinitely distant, and yet real, points of intersection), among which those of the pangeometrical sort are not the least. The Euclidean list of definitions, postulates and axioms is vitiated, not, or, at least, not only, by the fact that his lines of distinction between these several prerequisites of geometrical reasoning are not correctly drawn—that he confounds definitions with axioms and postulates with both,* and, besides, fails to discriminate between axioms of *quantity in general* and axioms of *spatial quantity*—but by his ignorance or disregard of the fact, to which I have referred, that every axiom, which is geometrically fertile, involves a definition.

lines can not inclose a space") are of the same nature with those of the 12th (usually called the 11th—the axiom of parallelism) has long been known. "La définition et les propriétés de la ligne droite," says D'Alembert (Élémens de Philosophie, § 12—Œuvres, tome i, p. 280) "ainsi que des lignes parallèles sont donc l'écueil et, pour ainsi dire, le scandale des élémens de géométrie."

* Hankel (Vorlesungen ueber die complexen Zahlen und ihre Functionen, p. 52) draws attention to the fact that the confusion above referred to is chargeable, not to Euclid, but to his editors and commentators. "In all the manuscripts," says Hankel, "which F. Peyrard has collected in preparing his excellent edition of Euclid (Œuvres d'Euclide trad. en Latin et en Français, tome i, p. 454) the famous 11th principle of the theory of parallels appears, not among the κοιναὶ ἔννοιαι relating to equal and unequal quantities, but as the 5th postulate (αἴτημα). Similarly the 10th axiom in all these manuscripts appears as the 4th postulate, while the MSS. vary in respect to the 12th axiom, it being thus evident that the three axioms owe the place which unaccountably they still occupy in the list of axioms to a misunderstanding."

And this ignorance—very excusable in Euclid's time—unfortunately appears to be shared by the writers of geometrical text-books at the present day.

One of the points upon which the debate between Helmholtz and his opponents has largely turned is the question whether or not Beltrami's pseudo spherical space is conceivable or imaginable (*vorstellbar*); and, to maintain the affirmative, Helmholtz propounds a remarkable definition of imaginability. He defines the power of imagining spatial forms as "the power of fully representing the sense-impressions which the object would excite in us according to the known laws of our sense-organs under all conceivable conditions of observation, and by which it would be distinguished from other similar objects." * Whatever may be the general merits of this definition, it is certainly obnoxious to the charge of irrelevancy. As the old logicians would say, it is founded upon an *ignoratio elenchi*, a misapprehension of the question. Granting, for the sake of argument, that the act of imagining a form of space is truly described as an anticipation of sense-impressions, the question as to the existence of the power sought to be defined is, not what would be the nature of these impressions, but whether or not they could coexist in the imagination in the required spatial order and form according to the known laws of the representative faculty. Helmholtz refers to the attempt of Beltrami to make pseudo spherical space representable by projecting its points, lines and surfaces upon the interior of an ordinary spherical surface "whose points correspond to the infinitely distant points of pseudo-spherical space," and claims that this attempt is successful. In the same sense Professor Sylvester, in the note to his Exeter

* "Origin and Meaning of Geometrical Axioms," Mind, vol. iii, p. 215.

address already quoted, observes that "every perspective representation of figured space of four dimensions is a figure in real space, and that the properties of figures admit of being studied to a great extent, if not completely, in their perspective representations." And it has become a standing assertion of the pangeometers that the forms of a space of any given dimension may be projected into the space of the next lower dimension. But this assertion, at best, holds good only for the limits of projection, where the resulting point or figure ceases to be an explicit reproduction of the figure projected. When a straight line is projected orthogonally upon another straight line at right angles to it, it appears as a point; a form of the first dimension is, in a sense, reduced to the dimension zero. But the representative point, by itself, does not enable us to reproduce and reason about the line whereof it is the projection. It may be said that we can at least know that the line projected is straight; but that is a conclusion which follows only from the properties of lines as they are otherwise known; from the mere inspection of the point it is not even inferable that it is a projection of a line at all. Similarly a plane may be projected upon another plane, so as to appear as a line, a form of two dimensions being reduced to a form of one dimension; but it does not follow that we may study the properties of the plane by merely contemplating or analyzing the line. The so-called projections of solids upon surfaces are in fact projections, upon a normal surface, of several surfaces making different angles with it, and the inferences from such a projection respecting the properties of geometrical solids depend upon our associations of visual with tactual impressions in which our apprehension of geometrical solidity has its origin. There being confessedly no tactual or other impressions evidencing the existence of a fourth dimension, the analogy upon which the alleged imaginability of transcendental space-forms is founded is without support.

But it is of little consequence what ground there is for the claim (which has recently been urged in another form by Felix Klein *) that the resources of projective geometry are sufficient to enable us to represent the properties of a space of more than three dimensions in tridimensional space: for the question of representability is wholly foreign to the matter in dispute. If it were shown, for instance, that a pseudo-spherical surface may be, mentally or really, traced *in* space, this certainly would not prove, or tend to prove, that space is inherently pseudo-spherical. There is no doubt about the imaginability of a spherical surface, but from this it does not follow that space itself is spherical. To support the conclusion of the immanent pseudo-sphericity of space it would be necessary to maintain that none but pseudo-spherical surfaces can exist, and, therefore (conformably to the teachings of sensationalism), be represented, or imagined as existing, in it. And, in view of this, the whole argument of Helmholtz not only ceases to be available as a support of geometrical transcendentalism, but recoils upon himself. If pseudo-spherical surfaces can be imagined to exist, and therefore, upon his own principles, are possible in "flat" space, why can not ordinary straight lines and flat surfaces exist in pseudo-spherical space? And what, then,

* "Ueber die Nicht-Euklidische Geometrie." Math. Ann., vol. iv, p. 573. In this article, as in nearly all the writings of the pangeometers, who deal with imaginary and infinitely distant points *ad libitum*, analytical representability (by means of symbols among which infinite and imaginary elements are treated as coördinate with real elements) is confounded with imaginability.

becomes of his telescopic test of the curvature of space? Or am I under a misapprehension as to Helmholtz's true meaning?—does he simply contend that pseudo-spherical surfaces would be imaginable by pseudo-spherical beings with pseudo-spherical organs of sense, and consequent pseudo-spherical intellects in a pseudo-spherical space, if it existed? That is a proposition which even Land and Krause would hardly dispute.

The history of cognition affords no illustration, perhaps, of the irrepressibility of intellectual traditions which is more instructive than the doctrines of transcendental geometry. Glancing back at the contents of the present chapter, we see that even the science of mathematics—the exactest of all the sciences, whose methods are said to be as infallible as its foundations are supposed to be permanent, and which, ever since the dawn of human intelligence, has pursued the even tenor of its way amid all the vicissitudes of speculation—is not exempt from the prepossessions of ontological realism. The same hypostasis or reification of concepts, which has given rise to the atomo-mechanical theory in physics, has led to the doctrine of pangeometry in mathematics. The hypostasis of space, by the mathematicians, is a strict analogue of the hypostasis of mass and motion by the physicists.

The full extent, however, to which the minds of contemporary mathematicians are bewildered by the false light of ontology can be brought into still clearer view by a further examination of the speculative background of transcendental geometry, as it appears in the famous essay of Riemann already referred to.

CHAPTER XIV.

METAGEOMETRICAL SPACE IN THE LIGHT OF MODERN ANALYSIS.—RIEMANN'S ESSAY.

The essay of Bernhard Riemann, "On the Hypotheses which lie at the Base of Geometry," owes its great celebrity to the fact that he was a mathematical analyst of the first order, one of the favorite pupils of Gauss, under the inspiration of whose teachings, if not at his suggestion, the essay was written—by whom, in fact, it was presented, in 1854, shortly before his (Gauss's) death to the philosophical faculty of Goettingen, and by whom its cardinal propositions were expressly indorsed as an exposition of his own speculative opinions. Every intelligent reader of this essay will agree with me, I think, that its intrinsic merit is not at all commensurate with the attention with which it was received and the interest with which it is still generally considered. Not only are its statements, both of the problem and of the proposed methods of solution, crude and confused, but they bear the impress throughout of Riemann's very imperfect acquaintance with the nature of logical processes and even with the import of logical terms. It is apparent, from the whole tenor of the essay, that its author was an utter stranger to the discussions respecting the nature of space which have been so vigorously carried on by the best thinkers of our time ever since the days of Kant, and that he was so little familiar with the history of logic as to be without the faintest suspicion of the manifold ambiguity of such terms as "concept" and "quantity," and of the necessity of their exact definition preliminary to an inquiry respecting the very foundations of human knowledge.*

The general argument of the essay is, that the nature of space is to be deduced from its concept; that the formation of such a concept of necessity involves its subsumption under a higher concept; that this higher concept is that of a "multiply extended quantity;" that, in order to determine how many kinds of space are possible, it is requisite to ascertain in how many ways quantity may be "multiply extended" (*mehrfach ausgedehnt*); and that, after the number of conceptually

* Riemann himself modestly apologizes for the philosophical shortcomings of his essay on the ground of his inexperience in philosophical matters. But the crudeness of his speculations affords a very striking illustration, in my judgment, of the well-known fact that exclusive devotion to the labors of the mathematical analyst has a tendency to develop certain special powers of the intellect at the expense of its general grasp and strength. Although Sir William Hamilton, no doubt, overstated the case against the mathematicians, I believe that his suggestions are not wholly unworthy of attention, and that there is force in the words of D'Alembert (referred to by Sir William Hamilton), which it is perhaps safest to quote in the original, without translation: "Il semble que les grands géomètres devraient être excellens métaphysiciens, au moins sur les objets dont ils s'occupent; cependant il s'en faut bien qu'ils le soient toujours. *La logique de quelques uns d'entre eux est renfermée dans leurs formules et ne s'étend pas au dela.* On peut les comparer à un homme qui aurait le sens de la vue contraire à celui du toucher, ou dans lequel le second de ces sens ne se perfectionnerait qu'aux dépens de l'autre. Ces mauvais métaphysiciens dans une science où il est si facile de ne le pas être, le seront à plus forte raison infailliblement, comme l'expérience le prouve, sur les matières ou ils n'auront pas le calcul pour guide. Ainsi la géométrie qui mesure les corps, peut servir en certains cas à mesurer les esprits même." D'Alembert, Élémens de Philosophie, § 11; Œuvres, tome i, p. 276.

possible varieties or species of multiple extension has thus been fixed, it is a matter of experimental determination which of these varieties or species is represented by our space, i. e., by the space in which the world, as we know it, has its being. After thus asserting that the concept "space" is to be subsumed under the concept "quantity," Riemann proceeds to declare that all quantities are in their nature multiples or aggregates (*Mannigfaltigkeiten*) which are *continuous* whenever there is continuous transition from one of their several "specializations" to the other, and *discrete*, when there is no such transition; that the "specializations" of discrete quantities are called *points*, and those of continuous quantities *elements*; and that continuous quantities are determined by measurement, while discrete quantities are determined by numeration. Space, according to Riemann, though a continuous quantity, is a quantity of *n*-fold (geometrical) extension, and is thus a Multiple or Aggregate, and therefore a quantity, notwithstanding its continuity. The degree of the multiplicity of this extension—i. e., the fact of its being simple, twofold, threefold, or generally *n*-fold—determines the (logical) extension of the concept space.

We have here five distinct propositions, which, for convenience of reference and discussion, may be stated in distinct form as follows:

1. That the nature of space is to be deduced from its concept.

2. That the concept of space can be formed and determined only by its subsumption under a higher concept.

3. That our space is a "triply extended Multiple or Aggregate," the higher concept under which its concept is to be subsumed being that of an "*n*-fold ex-

tended Multiple" or a "multiply extended Aggregate" (*eine n-fach ausgedehnte Mannigfaltigkeit*), and that—translating Riemann's phraseology into its plain logical import—the (logical) extension of this higher concept determines the number of the possible kinds of space.

4. That the conceptual possibility of space is coextensive with its empirical possibility, though not with its empirical reality.

5. That continuous quantities are coördinate with discrete quantities, i. e., are species of the same genus, both being in their nature multiples or aggregates.*

* The order and numeration of these propositions is, of course, my own; in Riemann's essay they appear in very promiscuous order. In proof of the general correctness of my statement of Riemann's doctrines, it is perhaps well to quote the introductory part of his essay in the original, italicizing the more important passages:

"*Ueber die Hypothesen welche der Geometrie zu Grunde liegen.*

"*Plan der Untersuchung.*

"Bekanntlich setzt die Geometrie sowohl den Begriff des Raumes, als die ersten Grundbegriffe fuer die Constructionen im Raume als etwas Gegebenes voraus. Sie giebt von ihnen nur Nominaldefinitionen, waehrend die wesentlichen Bestimmungen in Form von Axiomen auftreten. Das Verhaeltniss dieser Voraussetzungen bleibt dabei im Dunkeln; man sieht weder ob und in wie weit ihre Verbindung nothwendig, noch *a priori*, ob sie moeglich ist.

"Diese Dunkelheit wurde auch von Euklid bis Legendre, um den beruehmtesten neueren Bearbeiter der Geometrie zu nennen, weder von den Mathematikern, noch von den Philosophen, welche sich damit beschaeftigten, gehoben. Es hatte dies seinen Grund wohl darin, *dass der allgemeine Begriff mehrfach ausgedehnter Groessen, unter welchen die Raumgroessen enthalten sind, ganz unbearbeitet blieb. Ich habe mir daher zunaechst die Aufgabe gestellt, den Begriff einer mehrfach ausgedehnten Groesse aus allegemeinen Groessenbegriffen zu construiren.* Es wird daraus hervorgehen *dass eine mehrfach ausgedehnte Groesse verschiedener Maassverhaeltnisse fachig ist, und der Raum also nur einen besondern Fall einer dreifach ausgedehnten Groesse bildet.* Hiervon aber ist eine nothwendige Folge, dass die Saetze der Geometrie sich nicht aus allgemeinen Groessenbegriffen ableiten lassen, sondern dass diejenigen Eigenschaften, durch

I proceed to consider these propositions in their order.

1. The first proposition is in plain words a statement of the general ontological fallacy (discussed at length in

welche sich der Raum von andern denkbaren dreifach ausgedehnten Groessen unterscheidet, nur aus der Erfahrung entnommen werden koennen. Hieraus ensteht die Aufgabe, die einfachsten Thatsachen aufzusuchen, aus denen sich die Maassverhaeltnisse des Raumes bestimmen lassen—eine Aufgabe, die der Natur der Sache nach nicht voellig bestimmt ist; denn es lassen sich mehrere Systeme einfacher Thatsachen angeben, welche zur Bestimmung der Maassverhaeltnisse des Raumes hinreichen; am wichtigsten ist fuer den gegenwaertigen Zweck das von Euklid zu Grunde gelegte. *Diese Thatsachen sind, wie alle Thatsachen, nicht nothwendig, sondern nur von empirischer, Gewissheit, sie sind Hypothesen*, man kann also ihre Wahrscheinlichkeit, welche innerhalb der Grenzen der Beobachtung allerdings sehr gross ist, untersuchen, und hienach ueber die Zulaessigkeit ihrer Ausdehnung jenseits der Grenzen der Beobachtung sowohl nach der Seite des Unmessbargrossen, als nach der Seite des Unmessbarkleinen urtheilen.

"*I. Begriff einer n-fach ausgedehnten Groesse.*

"Indem ich nun von diesen Aufgaben zunaechst die erste, die Entwickelung des Begriffes mehrfach ausgedehnter Groessen, zu loesen versuche, glaube ich um so mehr auf eine nachsichtige Beurtheilung Anspruch machen zu duerfen, da ich in dergleichen Arbeiten philosophischer Natur, wo die Schwierigkeiten mehr in den Begriffen, als in den Constructionen liegen, wenig geuebt bin und ich ausser einigen ganz kurzen Andeutungen welche Herr Hofrath Gauss in der zweiten Abhandlung ueber die biquadratischen Reste, in den Goettingischen gelehrten Anzeigen, und in seiner Jubilaeumschrift darueber veroeffentlicht hat, und einigen philosophischen Untersuchungen Herbart's durchaus keine Vorarbeiten benutzen konnte.

"*Groessenbegriffe sind nur da moeglich, wo sich ein allgemeiner Begriff vorfindet, der verschiedene Bestimmungsweisen zulaesst. Je nachdem unter diesen Bestimmungsweisen von einer zu einer andern ein stetiger Uebergang stattfindet oder nicht, bilden sie eine stetige oder discrete Mannigfaltigkeit; die einzelnen Bestimmungsweisen heissen im ersten Fall Puncte, in letzterem Elemente dieser Mannigfaltigkeit.* Begriffe, deren Bestimmungsweisen eine discrete Mannigfaltigkeit bilden, sind so haeufig, dass sich fuer beliebig gegebene Dinge wenigstens in den gebildeteren Sprachen immer ein Begriff auffinden laesst, unter welchem sie enthalten sind (und die Mathematiker konnten daher in der Lehre von den discreten Groessen

the ninth chapter) that things and their properties are to be deduced from our concepts of them. As I have already said, Riemann does not define the term "concept;" nor does he inquire how concepts are formed or how they come to be possessions of the intellect. He says, indeed, that concepts of quantity are possible only when they can be subsumed under higher concepts,

unbedenklich von der Forderung ausgehen, gegebene Dinge als gleichartig zu betrachten), dagegen sind die Veranlassungen zur Bildung von Begriffen, deren Bestimmungsweisen eine stetige Mannigfaltigkeit bilden, im gemeinen Leben so selten, dass die Orte der Sinnengegenstaende und die Farben wohl die einzigen einfachen Begriffe sind, deren Bestimmungsweisen eine mehrfach ausgedehnte Mannigfaltigkeit bilden. Haeufigere Veranlassung zur Erzeugung und Ausbildung dieser Begriffe findet sich erst in der hoehern Mathematik.

"*Bestimmte, durch ein Merkmal oder eine Grenze unterschiedene Theile einer Mannigfaltigkeit heissen Quanta. Ihre Vergleichung der Quantitaet nach geschieht bei den discreten Groessen durch Zaehlung, bei den stetigen durch Messung.* . . . Fuer den gegenwaertigen Zweck genuegt es, *aus diesem allgemeinen Theile der Lehre von den ausgedehnten Groessen, wo weiter nichts vorausgesetzt wird, als was in dem Begriffe derselben enthalten ist,* zwei Puncte hervorzuheben, wovon der erste *die Erzeugung des Begriffs einer mehrfach ausgedehnten Mannigfaltigkeit, die zweite die Zurueckfuehrung der Ortsbestimmungen in einer gegebenen Mannigfaltigkeit auf Quantitaetsbestimmungen betrifft, und das wesentliche Kennzeichen einer n-fachen Ausdehnung deutlich machen wird.*"

I ought to say that my interpretations of several passages of this text are more or less conjectural. There is room for serious doubt, for instance, whether the expression "*Bestimmungsweisen*" is meant to denote the species comprehended by a genus, or the parts constituting a whole.—A wretched translation of Riemann's essay, which, by its clumsy literalism, materially adds to the obscurity and confusion of the original, was published in 1873, by W. K. Clifford (Nature, vol. viii, pp. 14 and 36 *seq.*). This translation was no doubt made, not *by*, but *for*, Professor Clifford, by some one who had a very insufficient knowledge of German. The merits of the translation are not unfairly instanced in the rendering of Riemann's term "*Mannigfaltigkeiten*" (varieties, multiplicities, used in the sense of multiples—Helmholtz translates "aggregates") by "manifoldnesses," of "*Groessenbegriffe*" by "magnitude-notions," etc. Of one passage the whole sense is changed by reading *koennten* for *konnten*.

or, as he expresses it, "when there is a general concept which admits of different specializations." But the question, where this process of subsumption begins or ends, and what are the nature and origin of the highest concept or *summum genus* of which all inferior genera or species must be specializations, does not occur to him. It is, however, an inevitable conclusion from Riemann's first proposition itself that he holds this most general concept to be an *a priori* form or possession of the mind, and that he believes the process of deduction by which its specializations are derived from it to be (in the language of Kant) a series of synthetic judgments *a priori*. In view of this a further consideration of the proposition is unnecessary; it is refuted by the whole tenor of the preceding chapters of this book. I may be permitted to observe, however, that it is without parallel in the entire history of intellectualism (usually called idealism); Kant, for example, expressly disclaims all belief in the doctrine that the intellect is aboriginally furnished with ready-made concepts.

2. The second proposition, that concepts of quantity can be formed and determined only by subsumption under more general concepts, is probably a vague reminiscence of the old logical rule that all definition is *per genus et differentiam*. In spite of Riemann's complaint, in the second sentence of his essay, that hitherto the science of geometry has given nominal definitions only of space and constructions in space—a complaint, by the way, which, so far as it applies to constructions in space, is unfounded—he does not seem to have a very clear insight into the nature of the distinction between definitions and concepts. For, if he had properly realized this distinction, he could not have failed to ask himself the question, what, under his definition,

became of the *summum genus* "quantity" which is the logical terminus of the processes of subsumption of which he speaks. Is this *summum genus* also a concept? Then it must be subsumable, in conformity with his rule, under a still higher concept, which, *ex vi termini*, it is not, being itself the highest. Or is it something else—a datum of experience? If it is, how then is the second proposition to be reconciled with the first, according to which everything is to be deduced from, as well as subsumed under, a concept? Or is this the old case of the hen in Newmarket which lays an egg, from which the same hen presently comes forth as a chicken?

The proposition here discussed almost at the outset involves our author in the most intolerable perplexity. "Concepts," he says, "whose specializations form a discrete aggregate (or multiple) are so common that, in the more cultivated languages at least, a concept may always be found under which things of whatever kind are subsumable." The meaning of this is, I take it, that of discrete aggregates there are always several similar or connatural kinds or species which may readily be subsumed under a higher concept. "But," he continues, "the occasions for the formation of concepts, whose specializations constitute a continuous aggregate, are so rare in ordinary life, that the places of things and colors are probably the only simple concepts whose specializations constitute a multiply extended aggregate"—that is to say, I suppose, there is but one species of a continuous aggregate or multiple other than space that admits of coördination and subsumption with it under the concept "multiply extended aggregate," viz., color. This singular statement (which, it may be noted parenthetically, is the exact reverse of the truth, there being, as we shall hereafter see, but one kind of discrete quantities, viz., numbers, and innumerable kinds of continuous quantities) has been elaborated with an extravagant expenditure of analytical power by Benno Erdmann,* who finds that there are *two* triply extended multiples which are coördinate and subsumable with space of three dimensions under the concept of a "continuous multiply extended aggregate:" sound and color. Sound, according to Erdmann, is a function of three independent variables, *acuteness*, *intensity*, and *timbre* (*Klangfarbe*). Similarly color depends on the variables *tone*, *degree of saturation* (*Saettigungsgrad*), and *intensity*.†

All this is simply puerile. To imagine that conclusions respecting the nature of space and the origin of its concept can be drawn from the mere fact that space is a function of three variables, and may thus in a manner be classified with similar functions, is a mockery of all reasoning from which an old scholastic would have turned with the scornful reminder that coördination and subsumption, for the purpose of effectually aiding in the formation of a particular concept, must not only be under a *genus*, but under a *genus proximum*.‡ Weissenborn's remark,# that on the same log-

* Die Axiome der Geometrie (Leipzig, 1877) p. 40 *seq.*

† It is significant, in this connection, that according to Helmholtz (who also falls in with Riemann's theory of conception) the three variables of the function "color" are the three primary colors of which each several color is said to be a mixture. "The Origin and Meaning," etc. Mind, vol. i, p. 309.

‡ Of this Erdmann seems to have some inkling, for he notes that space differs from color and sound in the circumstance of the absolute interchangeability of its three dimensions, the "dimensions" of color and sound not being interchangeable.

\# "Ueber die neueren Ansichten vom Raum," Vierteljahrsschrift fuer wissenschaftliche Philosophie, vol. ii, p. 321.

ical principles space might be coördinated with the amount of interest produced by a certain capital, which is a function of the three variables *capital*, *rate of interest*, and *time*, is perfectly just. And the number of species coördinate with space in the same sense might be indefinitely increased. For instance, space might be coördinated with the velocity of a railway-train on a straight road, inasmuch as this velocity is a function of the motive power of the engine, the weight of the train, and the grade of the track; or with the volatility of a liquid, which is a function of the nature of the liquid, its temperature, and the pressure of the atmosphere; or with the capacity of a man for labor, which depends on his general health and strength, the quantity of nourishment he has taken, and the amount of sleep he has had; and so on indefinitely. All this is very absurd, but not more so than the coördination of space with color and sound on the mere basis of the dependence of each on three variables which are arbitrarily called "dimensions."

3. I come now to Riemann's third proposition, that space is an "n-fold extended multiple" or a "multiply extended aggregate" (*eine mehrfach oder n-fach ausgedehnte Mannigfaltigkeit*"). The term "Mannigfaltigkeit," as here employed, is a standing puzzle to the readers of Riemann's essay. Weissenborn, who justly objects to the use of an adjective or predicative word in an appellative sense, for the denotation of a substantive entity, conjectures* that it was expressly devised by Riemann for the purpose of bringing the concept "space" within the scope of his second proposition. But this is a mistake. Riemann adopted the term from Gauss, who was probably the originator of its employ-

* *L. c.*, p. 320.

ment for the designation of "space in general" (as distinguished from "flat space," in the metageometrical sense).* Gauss, in turn, took the expression, no doubt, from Herbart,† to whose attempt at an elaboration of the idea of space from the manifold qualitative data of sense I have already referred, and whose philosophy is, to a great extent, a sort of reproduction of the old Eleatic quandaries about "The One and the Many." Herbart, in fine, had obtained it from Kant, whose disciple he was, or believed himself to be, and whose phrase "*Mannigfaltigkeiten der Empfindung*" is variously found, not only in his own writings, but also in those of his followers.

The only comment which I deem it necessary to make on this proposition is that space is not a "multiple" or "aggregate" at all, but that its very essence is continuity. This, as has been abundantly shown, follows from its conceptual nature as well as from its relativity. The determination of points in space, or "elements" of space, results from the establishment of quantitative relations between its parts, i. e., its purely arbitrary divisions, by means of numbers, in the manner to be considered presently. I have already shown, in the

* In his *Anzeige* of the Theoria residuorum biquadraticorum, Commentatio secunda, Gauss says: "Der Verfasser hat sich vorbehalten, den Gegenstand welcher in der vorliegenden Abhandlung eigentlich nur gelegentlich beruehrt ist, kuenftig vollstaendig zu bearbeiten, *wo dann auch die Frage, warum die Relationen zwischen Dingen, die eine Mannigfaltigkeit von mehr als zwei Dimensionen darbieten, nicht noch andere, in der allgemeinen Arithmetic zulaessige Arten von Groessen liefern koennen, ihre Beantwortung finden wird.*" Gauss, Werke, vol. ii, p. 178. This notice appeared originally in the Goettingische Gelehrte Anzeigen of April 23, 1831.

† In his Synechologie, e. g., Herbart speaks of "*die Mannigfaltigkeit der irrationalen Fortschreitungen in Bezug auf den Raum.*" Herbart's Werke, vol. iv, p. 153.

last chapter, that space itself is not, in any intelligible sense, a quantity.

4. Riemann's fourth proposition is founded on a confusion between *conceptual* possibility and *real* or *empirical* possibility. Conceptual possibility is determined solely by the consistency or inconsistency of the elements of the concept to be formed—it is tested simply by the logical law of non-contradiction; while empirical possibility depends upon the consistency of the thing conceived with the various conditions of sensible reality, or, what is the same thing, the laws of nature. This subject, also, has already been discussed to some extent in the last chapter, where it was pointed out that conceivability (in the strict sense of the term) of a thing or phenomenon is no proof of its imaginability or representability under the conditions of our physical and intellectual organization. Upon this distinction depend the utility and scope of the artifice not unfrequently resorted to, in certain analytical investigations, of supposing the existence of a fourth spatial dimension for the purpose of reducing certain functions to a symmetrical form; and this distinction, too, is the basis of an observation made by Boole * twenty-six years ago:

"Space is presented to us, in perception, as possessing the three dimensions of length, breadth, and depth. But in a large class of problems relating to the properties of curved surfaces, the rotation of solid bodies around axes, the vibration of elastic media, etc., this limitation appears in the analytical investigation to be of an arbitrary character, and, *if attention were paid to the processes of solution alone*, no reason could be discovered why space should not exist in four, or in any

* Laws of Thought, p. 175, note.

greater number of, dimensions. The intellectual procedure in the imaginary world thus suggested can be apprehended by the clearest light of analogy." Upon the same ground, and in the same sense, Hermann Grassmann, who is sometimes referred to as one of the founders of transcendental geometry, has developed the theory of extension in its general application to an indefinite number of dimensions, although he certainly did not cherish the delusion (as seems to be supposed by Victor Schlegel *) that this could be the source of inferences respecting the number of actual or empirically possible dimensions of space. On this subject we have Grassmann's own explicit declaration: † "It is clear," he says, "that the concept of space can in no wise be generated by thought. . . . Whoever maintains the contrary must undertake to derive the dimensions of space from the pure laws of thought—a problem which is at once seen to be impossible of solution."

5. Closely akin to his third and fourth propositions is Riemann's fifth proposition, that continuous quantities are coördinate with discrete quantities, both being in their nature multiples or aggregates, and therefore species of the same genus. This pernicious fallacy is one of the traditional errors current among mathematicians, and has been prolific of innumerable delusions. It is this error which has stood in the way of the formation of a rational, intelligible, and consistent theory of irrational and imaginary quantities, so called, and has shrouded the true principles of the doctrine of "complex numbers" and of the calculus of quaternions in an impenetrable haze.

The proposition that discrete and continuous quan-

* System der Raumlehre, preface, p. vi.

† Die lineare Ausdehnungslehre (1844) Einleitung, p. 20 *seq.*

tities are coördinate species of the same genus amounts to nothing less than the thesis that signs are logically coördinate with their significates. There are no "discrete quantities" except those which are dealt with in special (common) and general arithmetic, that is to say, *numbers*. Now, a number is an aggregate or collection of units each of which simply represents *an act of apprehension*, whatever may be the extent or nature of the object apprehended. If this object is designated as a *quantity*, a number is not a quantity at all, nor a measure of quantity, but simply an intellectual vehicle of quantities—a purely subjective instrumentality for their comparison and admeasurement. All the uncertainty and confusion which are characteristic of the numerous attempts to define and classify quantities are due to the ignorance or neglect of this elementary truth. Quantity has been defined as "that which is susceptible of augmentation, diminution and division," and as "the genus of which magnitude and multitude are the species;" or quantities have been first divided into *extensive* quantities (space) and *intensive* quantities (forces, colors, sounds, and all subjective affections), and the extensive quantities have then been subdivided into *continuous* and *discrete*. Now, the fact is that all objects of apprehension, including all data of sense, are *in themselves*, i. e., within the act of apprehension, essentially continuous. They become discrete only by being subjected, arbitrarily or necessarily, to several acts of apprehension, and by thus being severed into parts, or coördinated with other objects similarly apprehended into wholes. To say that a datum of sensation or of subjective feeling is in itself discrete is to assert that it is absolute, and to deny that quantity is essentially relative. And to maintain (with those who speak of positive, negative, fractional, irrational, imaginary, complex, linear, or directional numbers) that number may be continuous is to ignore the plainest and most unmistakable fact in all our intellectual operations, and to misinterpret all the teachings of the history of mathematics. Numbers, in themselves, being mere groups or series of *acts* of intellectual apprehension without reference to their contents, are not and can not be positive or negative, much less fractional, irrational, or imaginary. They can, indeed, be applied, not only to data of sensation and of subjective feeling, but also, by analogy, to relations between them, including relations established by the intellect. They can, therefore, stand, not only for things, but also for their actions and reactions and for the operations to which they are subjected. A number may represent motion in a given direction and in the direction opposite to it, thus becoming affected by the signs *plus* and *minus;* but these signs do not indicate any change in the nature of numbers, but merely a particularity in their application. Similarly numbers may represent ratios and assume the form of fractions; but the numbers do not thereby cease to be what they are, viz., units or collections of units, and therefore essentially integers. Fractions can be properly called numbers only in the sense that they point to the division, *not of the primary units expressive of the original acts of apprehension, but of the objects apprehended*, into subordinate units. Again: numbers may be signs of operations upon quantities that can not be successfully performed, such as the reduction of the diagonal and the side of a square to a common measure —in other words, the establishment of a definite numerical ratio between two quantities which do not admit of such a ratio. In such case the futility of the attempt finds expression in a sign prefixed to a number which, together with its significate, is ordinarily termed an irrational quantity; but the irrationality lies, not in the number, but in the attempt at its application to incommensurable magnitudes. The same thing is true, *mutatis mutandis*, of "imaginary quantities" and "complex numbers." The object of the act of apprehension, which is represented by the numerical unit, may be, not only rectilinear motion or transference in a given direction, but also angular motion; as the calculus of quaternions expresses it, the unit of operation may be a tensor, or a versor, or both; whence it follows that whenever the attempt is made to represent such an operation in terms of linear units with their positive or negative prefixes indicative of a fixed direction in which the motions, whereof the lines are the measures, occur, the attempt again fails, and this fact emerges in the form of the symbol which (being part of a system of symbolization that is not comprehensive enough to embrace the new operation) assumes a so-called imaginary form. But here once more, it is not the number which is imaginary, *but the operation as interpreted in conformity with the conventional rules of symbolization*, the consequence being that these rules have to be extended, and that the meaning of the symbols has to be widened. But this again imports a change, not in the nature of the signs, i. e., of the numbers, but in the nature and extent of their significates. In this manner the scope of arithmetical (and, of course, algebraic) symbolization is continually extended, not only by enlarging, but also by wholly changing the things, relations or operations which are successively the objects of intellectual apprehension. All this is perfectly safe and legitimate, provided that the change in the signification of the symbols be made in conformity with the logical canon of consistency, and with due regard, moreover, to the effect of such change upon the validity of the rules governing the syntheses and analyses to which the symbols are subjected. In the operation of ordinary arithmetical or algebraic multiplication, for instance, the law of commutation is of universal validity. Multiplication being nothing more than an abbreviated addition, the multiplicand and the multiplier may exchange places or functions without any effect upon the result. In the calculus of quaternions the mathematician generalizes the principle of multiplication, defining it as a process of finding a quantity which is produced from, or related to, the multiplicand in the same way in which the multiplier is produced from, or related to, the unit. Under this new definition he multiplies lines and other quantities into each other; but now it appears that the law of commutation is no longer generally applicable. The reason is that the apparent expansion of the principle of multiplication was in fact also a limitation, or rather a shifting of the meaning of the arithmetical or algebraic symbol—a removal of the condition upon which the validity of the law of commutation depended. I may observe here, incidentally, that it is a mistake to say, with Kelland and others, that the calculus of quaternions grows out of the common arithmetical or algebraic calculus by the removal of limitations. The example just adduced shows that it may involve an imposition of limitations as well. For this reason Peacock's law, which he calls the "principle of the permanence of equivalent forms,* viz., that "whatever algebraical forms are equivalent, when the symbols are general in form but specific in value, will be equivalent likewise

* Peacock, Symbolical Algebra, p. 59.

when the symbols are general in value as well as in forms," in order to be available as the fundamental principle of the theory of "complex numbers," requires a modification far more serious than is implied in Hankel's new statement of it as "the principle of the permanence of formal laws." For the expression "formal laws" is ambiguous and leaves us in doubt as to what laws are formal in the sense of being applicable to all the operations which are in any way representable by arithmetical or algebraic symbols.

The error respecting the true nature and function of arithmetical and algebraic quantities has become next to ineradicable by reason of the inveterate use of the word "quantity" for the purpose of designating indiscriminately both extended objects or forms of extension and the abstract numerical units or aggregates by means of which their metrical relations are determined. The effect of this indiscriminate use is another illustration of the well-known fact in the history of cognition that words react powerfully upon the thoughts of men, and by this reaction become productive of incalculable error and confusion. It is not to be expected, of course, that mathematicians will cease, at this late day, to speak of arithmetical or algebraic symbols as "quantities;" but there may be a little hope for the suggestion that they might return to the old phrase "geometrical (and other) magnitudes." The mischief lies, not so much in the use of a particular word, as in the employment of the same word for the denotation of objects differing from each other *toto genere*.*

* The perplexities occasioned by the use of improper and misleading terms in mathematics are animadverted on by Gauss himself in the notice already cited (Werke, vol. ii, p. 178), where he speaks of the obscurity incident to the interpretation of "negative and imaginary numbers," and

The ignorance or oblivion of the distinction here referred to also illustrates a phase in the history of error exemplifications of which have repeatedly been met with in the preceding pages: the confusion between purely conventional forms of thought and speech and forms or laws of objective existence. This confusion, which is at the bottom of the old assumption that our arbitrary or conventional classifications of natural phenomena are coincident with essential distinctions between them and can be used as a source of inferences respecting their nature and origin—that, as some one has said, the score of the Lord's creation, like that of Haydn's Creation, is crossed with bars—has been prolific of an endless train of fanciful presumptions by which the progress of science is incessantly obstructed.

For the reasons here set forth, the terms "abstract and concrete numbers" are also fallacious and misleading. Numbers, in themselves, are essentially abstract. In another sense they are necessarily concrete: they always stand for some particular object, relation, or operation. They are nothing in themselves. This remark is doubly true of algebraic symbols which require interpretation, in the first place, by assigning to them particular numerical values, these, in turn, remaining without significance until the units, of which they consist, are referred to their proper objects, relations, or operations. This is, no doubt, Duehring's meaning when he observes, somewhere in his History of the Principles of Mechanics, that algebraic symbolization is radically defective inasmuch as it makes no display of the numerical units which are the essential coefficients

observes: "If $+1$, -1, $\sqrt{-1}$ had not been called positive, negative, imaginary (or even impossible) units, but, for example, *direct*, *inverse*, *lateral* units, this obscurity would have vanished."

of every literal symbol. He might have extended this observation by adding that the use of letters as algebraic symbols, i. e., as representatives of numbers, is in itself a serious (though, perhaps, an unavoidable) infirmity of mathematical notation. In the simple formula, for instance, expressive of the velocity of a moving body in terms of space and time $\left(v = \frac{s}{t}\right)$, the letters have a tendency to suggest to the mathematician that he has before him direct representatives of the things or elements with which he deals, and not merely of their ratios expressible in numbers. In every algebraic operation the use of letters obscures the real nature, both of the processes and of the results, and tends to strengthen ontological prepossessions.

The true theory of the relations between arithmetical or algebraic quantities and magnitudes of extension was stated long ago, in Germany by Martin Ohm and in England by George Peacock (the Dean of Ely), Augustus de Morgan, D. F. Gregory, and others; but the writings of these thinkers have produced little impression upon contemporary and succeeding generations of mathematicians. This is peculiarly apparent in the books and articles expository of the theories of "imaginary quantities" and "complex numbers," and of the doctrines of the calculus of quaternions. The immense extension of the sphere of analysis since Descartes's new application of algebra to the determination of geometrical magnitudes is almost universally attributed to a growing insight into the true character of "arithmetical quantities," and to a progressive explication of the essential implications of number. It is supposed that Euclid's denial of the existence of numerical ratios between incommensurable quantities, as well as the protests of the early occidental arithmeticians and algebraists against negative or irrational numbers as "*numeri absurdi infra nil*" or "*numeri ficti*," or the designation by Girolamo Cardano of the negative roots of an equation as "*æstimationes fictæ*" representing solutions "*verè sophisticæ*," are one and all simply evidences of the ignorance of these several writers of the real nature of numbers. It is not at all unusual to meet with the dogma, in treatises on the theory of "complex numbers," that algebra and arithmetic are essentially linear, numeration being impossible except by progression, in equal steps, in the direction of a straight line.* And, I may add, the belief is by no means uncommon that metageometry is an advance beyond the old doctrines concerning the relations between geometrical forms in ordinary space, in the same sense and by the same logic in and by which the calculus of quaternions is an advance beyond ordinary analytical geometry.

The foregoing discussion has brought us to the point where the reader is in a condition, I hope, to realize the great fundamental absurdity of Riemann's endeavor to draw inferences respecting the nature of space and the extension of its concept from algebraic representations of "multiplicities." An algebraic multiple and a spatial magnitude are totally disparate. That no conclusions about forms of extension or spatial magnitudes are derivable from the forms of algebraic functions is evident upon the most elementary considerations. The same algebraic formula may stand for the most various things. Equations of the second degree, for example, may represent either geometrical areas, or geometrical curves. The equation $y = x^2$ may represent, either the area of a square whose side is x, or a parabola (referred

* Cf. Riecke, die Rechnung mit Richtungszahlen (Stuttgart, 1856).

to an axis of ordinates) whose parameter is 1. If Riemann's argument were fundamentally valid, it could be presented in very succinct and simple form. It would be nothing more than a suggestion that, because algebraic quantities of the first, second, and third degrees denote geometrical magnitudes of one, two, and three dimensions respectively, there must be geometrical magnitudes of four, five, six, etc., dimensions corresponding to algebraic quantities of the fourth, fifth, sixth, etc., degree.*

It is hardly necessary to say, after all this, that the analytical argument in favor of the existence, or possibility, of transcendental space is another flagrant instance of the reification of concepts.

* It is not unworthy of remark, here, that the practice of reading x^2 and x^3 as x square and x cube, instead of x of the second or third power, is founded upon the silent or express assumption that an algebraic quantity has an inherent geometric import. The practice is, therefore, misleading, and ought to be disused. *Principiis obsta!*

CHAPTER XV.

COSMOLOGICAL AND COSMOGENETIC SPECULATIONS.—THE NEBULAR HYPOTHESIS.

LIKE all metaphysical theories, the atomo-mechanical theory has its cosmogonies. All metaphysical cosmogonies are attempts to deduce the universe and its phenomena from one or more primordial elements by the application of a few general principles. The cosmogonies of the atomo-mechanical theory are attempts to deduce the universe and its phenomena from the elements of mass and motion by the application of mechanical principles expressive of the simple laws of motion. As has been shown, the ultimate problem of the atomo-mechanical theory, to whose effectual and complete solution the physicists of the day look forward with a greater or less degree of confidence—though many of them are clear-sighted enough to regard it as an aspiration never to be realized—is the exhibition of all vital and organic phenomena as results of ordinary chemical and physical action, and of chemical and physical action, in turn, as exchanges and transferences of mechanical motion between constant and uniform elements of mass.

A question necessarily preliminary to cosmological speculations of whatever kind has been extensively mooted, of late, by mathematicians and physicists alike —the question respecting the finitude or infinitude of the universe in time, space, and mass.* A cosmogony, properly so called, inevitably involves the presumption that the universe is finite in past time at least, for it is a theory respecting the origin or *beginning* of the universe. The vision of the cosmogenetic theorist extends backward, either to the absolute nothing, or to a state of physical uniformity wholly destitute of those phenomenal differences and changes which are the essential prerequisites of the notion of time. This universal cosmogenetic presumption of the finite duration of the universe in the past has recently been supplemented by the assertion of its limited duration in the future—an assertion founded on a variety of physical considerations, the most noteworthy among which is the doctrine of the progressive dissipation of energy. This doctrine is stated in the most intelligible form, perhaps, by Sir William Thomson,† and is embodied in the following propositions:

"1. There is at present in the material world a universal tendency to the dissipation of mechanical energy.

"2. Any restoration of mechanical energy, without more than an equivalent of dissipation, is impossible in inanimate material processes, and is probably never effected by material masses either endowed with vegetable life, or subjected to the will of an animated creature.

"3. Within a finite time past the earth must have been, and within a finite period of time to come the earth must again be, unfit for the habitation of man as

* Cf. Wundt, "Ueber das Kosmologische Problem," Vierteljahrsschrift fuer wissenschaftliche Philosophie, vol. i, p. 80 *seq.*

† "On a Universal Tendency in Nature to the Dissipation of Mechanical Energy," Phil. Mag., series iv, vol. iv, p. 304 *seq.*

at present constituted, unless operations have been, or are to be, performed which are impossible under the laws to which the known operations going on at present in the material world are subject."

The reasoning by which these conclusions (which, it may be noted in passing, are carefully and in terms confined to our planet, or, at least, our planetary system) are arrived at is that, inasmuch as all the operations of nature, which constitute its life and action, depend upon transformations of energy, and as every such transformation, in conformity with the second law of thermo-dynamics, is in effect (to use the expression of P. G. Tait) a degradation from a plane of higher to one of lower transformability or availability, the ultimate effect must be a conversion of all the energy of the world into heat and a reduction of its temperature to absolute uniformity. From this state of uniformity in the diffusion of heat no restoration of available energy is possible; for heat admits of transformation into other forms of energy only by passing from a body of higher to one of lower temperature.*

* The doctrine of the dissipation of energy has been extensively developed by Clausius, who designates the sum of the possible transformations of the world's energy as its *entropy*, and announces that "the entropy of the world tends to a maximum." (Pogg. Ann., vol. cxxi, p. 1; Abhandlungen ueber die mechanische Waermetheorie, vol. ii, p. 44.) It is to be regretted that Tait, while adopting the word "entropy," undertakes to use it, as he himself says (Thermo-dynamics, § 48; *ib.*, § 178), "in the opposite sense to that in which Clausius employed it," and that Maxwell (Theory of Heat, pp. 186, 188) follows him. Nothing is more to be reprobated than an arbitrary change in scientific terminology, and especially a deliberate tampering with the received meaning of a term. It ought to be added that Tait does not even succeed in his attempt to reverse Clausius's meaning, and that Maxwell, too, is in error when he says that "Clausius uses the word (entropy) to denote the part of energy which is not available."

It is clear that, if the law of the dissipation of energy applies to the universe at large—that is to say, if the dynamics of a finite material system can be legitimately extended to the Cosmos as an infinite whole—there must, sooner or later, be an end of the universe identical with its beginning as assumed by the atomo-mechanical theory. The processes of nature must eventuate in a thorough homogeneity of its elements—in a complete absence of the differences and changes which constitute the attestation of its real or actual existence. This conclusion has been sought to be avoided by the assumption of the finitude of the universe in mass, or in space, or in both. The first impulse in this direction probably came from an article of W. M. Rankine* (published shortly after the appearance of that of Sir William Thomson), in which it was argued that "if there is between the atmospheres of the heavenly bodies an interstellar medium perfectly transparent and diathermanous—i. e., incapable of converting light and heat from the radiant into the fixed or conductible form, and thus incapable of acquiring any temperature whatever—and if this interstellar medium has bounds beyond which there is empty space, the radiant heat of the world will be totally reflected and will ultimately be reconcentrated into foci in which a star (i. e., an extinct mass of inert compounds) would be vaporized and resolved into its elements, a store of chemical force being thus reproduced at the expense of a corresponding amount of radiant heat."

The supposition of the finitude of the mass of the universe was not new; it had often been made before. But here it presented itself in a new form. Hitherto

* "On the Reconcentration of the Mechanical Energy of the Universe," Phil. Mag. (iv), vol. iv, p. 358 *seq.*

the supposition had been that the mass, though limited, was diffused throughout unlimited space; and in this form it has recently been revived by Wundt, who imagines that the finitude of a mass may be reconciled with the infinitude of its volume by the assumption of an endlessly progressive increase of its tenuity, the mass being taken as the finite sum of an infinite converging series. Rankine, on the contrary, required the physicist to grant that the mass of the universe is finite also in extent and is everywhere surrounded by void space. The conception of a material universe thus bounded in boundless space obviously presents insurmountable difficulties; and in view of these difficulties many astronomers and physicists hailed with delight the thesis of the metageometers that space itself, though unlimited by reason of its inherent curvature, is not infinite, and that, therefore, the mass of the universe must be finite, however diffused. This thesis was doubly welcome because it appeared, at first sight, also to afford the means of escape from another difficulty raised by the astronomers. In 1826 Olbers * observed that, if the number of bodies in the universe radiating heat and light is infinite, each point in space must receive an infinite number of caloric and luminar rays, and must, therefore, be infinitely hot and bright—adding, however, that this consequence could be avoided by supposing an absorption of the greater part of these rays by the dark and cold bodies in space. But this salvo at once appeared questionable, on the reflection that the dark and cold bodies disseminated among the luminous stars must speedily reach the point of incandescence, and that their absorbing power must soon be exhausted.

* Bode's astron. Jahrbuch, 1826, p. 110 *seq.* Quoted by Zoellner.

There is supposed to be a still further and similar perplexity, growing out of the fact of gravitation, especially in view of its instantaneous action. It is said that a universe consisting of an infinite number of bodies attracting each other would not only be without a definite center of gravity to which all cosmical motions could be referred—its center of attraction being everywhere, and therefore nowhere—but would result in an infinite pressure (I follow the expression of Wundt, though it would, perhaps, be more correct to say an infinite *strain*) at every point in space. This difficulty, in particular, is urged by Wundt as insuperable so long as the mass of the universe is held to be infinite; it can, in his opinion, be overcome only by the assumption that this mass is limited.

It is unnecessary to enter upon a minute examination of the validity of these considerations adduced in support of the theory of the finitude of the material universe. As to the last of them, relating to the effects of radiation and gravitation, it is readily seen, and has been pointed out by Lasswitz,* that they lose their force the moment we recollect that the intensity, both of radiation and gravity, decreases as the square of the distance increases, and that the infinite series expressive of the several effects of heat, light, and gravitation are converging, their summation yielding finite results. And of the application of the doctrine of the dissipation of energy to an infinite universe it is to be said that it is wholly inadmissible. That doctrine is, no doubt, irrecusable in its application to any finite material system. Every such system must come to an end, as it has had a beginning. And this is true of every such system, whatever its extent. But it is not true of

* Vierteljahrsschrift f. w. P., vol. i, p. 329 *seq.*

a universe absolutely unlimited. Neither the law of the conservation of energy, nor that of its dissipation, can be legitimately applied to it. The universe, taken as absolutely infinite, is not a conservative system and is not in any proper sense subject to physical laws. We can not deal with the Infinite as with a physically real thing, because definite physical reality is coextensive with action and reaction; and physical laws can not be applied to it, because they are determinations of the modes of interaction between distinct, finite bodies. The universe, so called, is not a distinct body, and there are no bodies without it with which it could interact. Operations with the term Infinite in analogy to operations with finite terms are as illegitimate in physics as they are in mathematics. The Infinite is simply the expression of the essential relativity of all material things and their properties, and is thus, in a sense, inherent in every finite form. It is the basis of all the relations which constitute sensible actuality, but it is not itself a group of such relations. It is the background of all material actions and forms; no system of elements or forces can exist without it, or is cognizable without reference to it; and in this sense, and in this sense only, the universe is necessarily infinite in mass as well as in space and in time.

It follows that all cosmogonies which purport to be theories of the origin of the universe as an absolute whole, in the light of physical or dynamical laws, are fundamentally absurd. The only question to which a series or group of phenomena gives legitimate rise relates to their filiation and interdependence; and the attempts to transcend the bases of this filiation and interdependence—to determine the conditions of the emergence of physical phenomena beyond the bounds of space and the limits of time—are as futile as (to use the happy simile of Sir William Hamilton) the attempt of the eagle to outsoar the atmosphere in which he floats and by which alone he may be supported.

This leads me to a discussion of a cosmogenetic theory which has attained to great celebrity and very general acceptance, under the name of the Nebular Hypothesis. As now generally held, this theory may be briefly stated as follows:

Primordially the materials, which are at present found, partly at least, conglomerated in the bodies composing the stellar, solar, planetary, satellitic, and meteoric systems, were uniformly dispersed throughout space. In some way, by the action of cosmic (attractive and other) forces, this uniformly diffused and very attenuated matter came to be divided into large nebulous spheres which began slowly to rotate, the rotation resulting, perhaps, from the act of division, or from internal differences in their densities and irregularities in their forms, which deflected the lines of gravitation from a strictly radial direction, the centers of attraction no longer coinciding with the centers of figure. In proportion as these spheres parted with their heat they contracted; and this contraction led to an increase of their velocities of rotation in conformity to a mechanical law known as the law of the *conservation of areas or of angular momentum.* This law, in its most general expression, is simply a corollary from the law of inertia, from which it follows that the resultant angular momentum of any material system can not be changed, either in magnitude or the direction of its axis, by the mutual action of its constituents.* For the purpose of

* All mechanical or dynamical laws of conservation—the conservation of momentum, of angular momentum and of energy—are (as I have

its application to a rotating nebulous mass, however, the law may be more intelligibly stated in another form, viz., that, whatever change of volume or form may be produced in a material system by the mutual attraction of its constituent elements, the sum of all the areas described by the *radii vectores* of the several elements or particles round the center of rotation, in a unit of time, is constant. Now, the areas being proportional to the squares of the diameters, it follows that the angular velocity increased with great rapidity as the contraction of a nebulous mass proceeded. An immediate consequence of this increase of velocity was a proportionate increase of the centrifugal force in the equatorial regions of the rotating sphere, so that in course of time this force came to balance, and afterward to exceed, the centripetal gravitation. This led at first to a disproportionate contraction of the sphere at the poles and to the assumption, by the sphere, of an oblately spheroidal or lenticular form, and eventually to successive detachments of equatorial rings or zones which at first circulated round the residual mass in the direction of its original rotation, but which—by reason of the instability of such rings in case of the least departure from absolute regularity of form or constitution—broke up into parts, forming one or more minor spheres or spheroids. These continued to revolve round the sun with a velocity nearly equal to the rotatory velocity of their mate-

already indicated in the sixth chapter) at bottom nothing more than applications of the principle of inertia to complex material systems. It is the great merit of Poinsot to have brought to light the formal analogies (prefigured, to a certain extent, in the writings of Euler) between the laws governing movements of rotation and those determining the forms of ordinary translatory motion. It is hardly necessary to add that the law of the conservation of areas is in form a generalization of Kepler's second law.

rials at the moment of their detachment and conglobation. In most cases, probably, the whole mass of such a ring coalesced into a single body, i. e., into a planet, while in some cases several bodies were formed, such as they appear in our planetary system in the zone of asteroids. Each of the planets, while revolving round the residual mass ,whose condensation is supposed to have produced the sun, also began to rotate on an axis of its own, the direction of this rotation coinciding with that of its revolution. It thus became subject to the same dynamical conditions which determined the evolution of the parent system; it also threw off rings which either retained their form (as in the case of the Saturnian rings) or formed into minor satellitic bodies.

The arguments which have been advanced in support of this hypothesis are so well known that it is hardly necessary to recapitulate them. Among them are the existence, in the stellar regions, of nebulous masses in various stages of condensation; the evidences of the increase of temperature from the surface of our planet toward the interior; the proximate coincidence of the orbital motions of the several planets, both in direction and plane, and the further proximate coincidence of this orbital motion with the direction and plane of the sun's rotation; the similar coincidence of the directions of the orbital motions of the satellites with the axial motions of their planets; the oblately spheroidal form of the earth, and, as far as we know, of the other planets, which peculiar form has not only been theoretically demonstrated, but has also been experimentally shown, by M. Plateau, to be the form necessarily assumed by a rotating body in a liquid or semi-liquid state. These considerations were adduced, almost in the same order and form, by Kant and Laplace, and they have since been supplemented by a variety of other considerations more or less plausible, among which may be mentioned the agreement of the theoretical consequences of the fact, that the projection of planetary masses from the parent globe must have taken place with ever-increasing rapidity as the contraction of the globe progressed, with certain well-known features of our own planetary system. Attempts not wholly unsuccessful have even been made to effect a deduction, from the elements of this theory, of the empirical law respecting the distances of the several planets from the sun which is known as the law of Bode or Titius.

The nebular hypothesis, as a theory of the origin, not only of our planetary system, but of stellar and planetary systems throughout the universe, is commonly ascribed to Laplace, who is supposed to have been unaware of the fact that the hypothesis which he advanced had been published by Kant, in his *Naturgeschichte des Himmels*, in 1755, nearly half a century before the first appearance of the *Exposition du Système du Monde*, in 1796. But the truth is that the Nebular Hypothesis, in the form in which it is now generally held, is due to Kant, and differs in several essential particulars from the hypothesis of Laplace. This latter hypothesis is limited in terms to our planetary system, and there is no indication in any of the writings of the French astronomer—certainly none in his *Exposition du Système du Monde*—that he ventured to extend it to the entire universe, as was expressly done by Kant. But there is a difference still more important between the hypotheses of the two thinkers. Kant's assumption was that "*all the materials composing the spheres* that belong to our solar world were, in the beginning of all things, resolved into their elementary substance and filled the whole space of the system in which these spheres now move." * This assumption is common to all recent forms of the nebular hypothesis that have fallen under my notice—they all postulate a diffusion of the entire mass of the sun, planets, comets, and satellites constituting our planetary system throughout the planetary space. The assumption of Laplace, on the contrary, is simply that *the atmosphere of the sun* at one time extended beyond the orbits of the farthest planets, and that the formation of the planets and their satellites as well as that of the comets was due to a gradual cooling and contraction of this atmosphere.†

It is hardly necessary to say that the Laplacean form of the nebular hypothesis is far too narrow to serve the purposes of a general cosmological theory. Such a theory demands the derivation of the several concretions of cosmical matter from some primitive homogeneous mass. This demand is complied with by the hypothesis of Kant; but it is very partially, if at all, satisfied by that of Laplace. And this brings us into the presence of a formidable difficulty. It is to be feared that, in proportion to its amplification to cosmogenetic dimensions, the nebular hypothesis parts with its validity as a physical theory. This subject was ex-

* "Ich nehme an, dass *alle Materie, daraus die Kugeln die zu unserer Sonnenwelt gehoeren, alle Planeten und Kometen bestehen*, im Anfang aller Dinge in ihren elementarischen Grundstoff aufgeloes't, den ganzen Raum des Weltgebaendes erfuellt haben, darin jetzt diese gebildeten Koerper herumlaufen." "Naturgeschichte des Himmels," Kant's Werke, vol. vi, p. 95.

† "La considération des mouvemens planétaires nous conduit donc à penser qu'en vertu d'une chaleur excessive *l'atmosphère du soleil* s'est primitivement étendue au delà des orbes de toutes les planètes, et qu'elle s'est resserrée successivement jusqu'à ses limites actuelles." Système du Monde (2me éd.), p. 345.

amined, nearly twenty years ago, by M. Babinet, in an article on the Cosmogony of Laplace,* in which he shows that the actual rotatory velocities of the several planets are in fact vastly greater than the velocities to be deduced, by the aid of the law of the conservation of areas, from the nebular hypothesis, if that hypothesis includes the assumption of a diffusion of the solar mass itself throughout a space coextensive with the limits of our planetary system. "Several persons," says M. Babinet, "have thought that the sun himself had originally been expanded so as to fill the entire space now occupied by the planets, although Laplace expressly mentions that at the moment of the formation of these bodies it was only *the atmosphere* of the sun which had this vast extent. We are able to test this question mathematically, by calculating from the sun's actual period of rotation, which is twenty-five and three tenths days, what would be the velocity of rotation if, conserving the sum of the areas described by all its material points, it were expanded so that its radius, which is now equal to one hundred and twelve times the equatorial radius of the earth, became equal to the distance from the earth to the sun, or from Neptune to the sun. . . . The calculation on the first of these bases gives a rotation of 1,162,000 days, amounting to more than three thousand (3,181) years. The period of revolution calculated on the second basis would evidently be nine hundred times greater, that is to say, more than twenty-seven thousand centuries.

* "Note sur un Point de la Cosmogonie de Laplace," Comptes Rendus, vol. lii, p. 481 *seq.* My attention was drawn to this article by a passage in an interesting little pamphlet of Dr. E. Budde, of Bonn, Zur Kosmologie der Gegenwart (Bonn, ed. Weber, 1872), to which I shall have occasion to recur hereafter.

"These numbers being infinitely greater than those expressive of the earth's and Neptune's actual periods of revolution, it is plainly impossible to admit that these planets have been formed out of the solar mass itself extended beyond the planetary orbits. This, however, does not preclude the idea that the stars themselves have been formed at the expense of a universal cosmic matter endowed with excessively feeble movements of rotation round the center of gravity of each mass which was in process of formation as an independent sun.

"The conclusion is that, if the entire mass of the sun had been expanded to the limits of the planetary system, it must have had a movement of rotation far too feeble to enable the centrifugal force to balance the force of gravity so as to lead to the separation of an equatorial ring from the total mass."

The discrepancies here brought to light between the actual orbital periods of the planets and the corresponding periods found by calculation in accordance with the postulates of the nebular hypothesis, are so enormous that there appears to be no possibility of accounting for them by the assumption of a progressive contraction of the orbits of the several planets since their projection, and the consequent quickening of their orbital motions.

The calculations of M. Babinet do not constitute the only difficulty which besets the nebular hypothesis, either in its general cosmogenetic or in its special Laplacean form. In the progress of astronomical discovery it has appeared that several of the supposed coincidences between the facts and the hypothesis fail. Thus, there appears to be an exception to the directional uniformity of the axial and orbital motions of the planets and their satellites in the case of Uranus, the orbital planes of whose satellites are nearly perpendicular to the ecliptic, the circumplanetary motions of the satellite as well as the axial motion of the planet, moreover, being retrograde—a fact long since discovered by Sir William Herschel, and confirmed by various subsequent observations. Another difficulty has been suggested by the recent discovery (1877), by Professor Asaph Hall, of two satellites of the planet Mars, and the proximate determination of their respective distances from the primary, as well as their orbital (circumplanetary) periods. It was found that the distances of the inner and outer satellites from the center of the planet are about three and six times, respectively, the radius of the planet, and that the periods of revolution of these satellites are 7·65 and 30·25 hours, respectively, while the period of rotation of the planet (Mars) itself is 24·623 hours. It appears, then, that one of the satellites revolves about the planet in less than one-third of the time required for the planet's axial rotation.

At first sight this fact would seem to be wholly incompatible with the nebular hypothesis. In the light of this hypothesis, the orbital motions of a satellite are continuations of the axial motions of the materials out of which the satellites are formed; its orbital period ought, therefore, to be equal, proximately at least, to the period in which the planet rotated at the time of the satellite's formation. And that period is of necessity greater than the period of the planet's present rotation, by reason of the acceleration produced by its subsequent contraction.

Two attempts have been made, thus far, to reconcile the anomaly just referred to with the postulates of the nebular hypothesis. One of these is based on the supposition that the orbits of the satellites have been contracted by the resistance of the ethereal medium which was formerly supposed to have shortened the period of Encke's comet. But this resistance is wholly insufficient to account for the anomaly, even if the very doubtful existence of an interstellar and interplanetary medium capable of offering material resistance to planetary motion be granted. The other attempt seeks to refer the anomaly to a retardation of the planet's period of rotation, and a corresponding quickening of the revolving motion of the satellites by tidal action. While it is admitted that the retardation of the rotatory period of the planet by the action of the tides raised upon it by a satellite could at most produce a coincidence of that period with the orbital period of the satellite, it is claimed by G. H. Darwin that the period of planetary rotation may be lengthened beyond that of satellitic revolution by the friction of the tides due to solar action. The effect here alluded to is the result of a conversion of the energy of planetary rotation into heat, and of a transference of the angular momentum of rotation to that of orbital revolution of the bodies by whose mutual attraction the tides are raised about their common center of inertia; and, inasmuch as a larger planet has both more energy of rotation and more angular momentum than a small one, the most rapid changes in the periods of rotation and revolution will be produced in the case of the smallest planet. Mars being the smallest planet attended by satellites, it is said that the slowness of his rotation as compared with the revolution of his inner satellite may have been produced by the agency above indicated within the period assigned to the history of our planetary system by the advocates of the nebular hypothesis.

Whatever may be the weight of the several objections urged against certain features, more or less essential, of the nebular hypothesis, there is one objection which is fundamental: the inadmissibility, already pointed out, of all speculations respecting the origin of the universe as an unlimited whole. But, apart from this, it is plain that the derivation of the forms and movements of the stellar and planetary systems from a primordial homogeneous mass uniformly diffused throughout space is impossible. In the first place, such a mass must be either at rest or in uniform motion; and this state of rest or uniform motion, according to the most elementary principles, could be changed only by extraneous impulses or attractions. And, there being no "without" to the all-embracing Cosmos or Chaos, the original state of rest or uniform motion would necessarily be perpetual.* In the second place, such a nebulous universe would be of perfectly uniform temperature; all parts would be equally hot (or cold), and there could be no radiation or loss of heat resulting in a contraction of any part of the nebulous mass. Its thermo-dynamical condition would be constant for the same reason which establishes the permanence of its general dynamical condition.

The cumulation of difficulties presented by the nebular hypothesis has become so great, and is beginning to be so extensively realized, as to develop a tendency to modify or supplant it by another hypothesis which may be called the hypothesis of meteoric agglomeration. This hypothesis commends itself to the modern physicist by reason of its apparent exemplification of the general doctrine that, for the purpose of ascertaining the nature of the agencies which have produced

* As Duehring expresses it (Kritische Geschichte der allgemeinen Principien der Mechanik, 2d ed., § 151), "If ever there had been perfect equilibrium between the parts (of the nebulous mass) it would continue to exist now."

a particular physical system or form, we must in the first instance look to the agencies concerned in its maintenance or destruction—a doctrine which might be condensed into a canon: *quod sustinet vel delet, formavit.* This doctrine is in effect nothing more than a new statement of the old law of parsimony which forbids the unnecessary multiplication of explanatory elements and agencies. It has been extensively and successfully applied in geology, which now endeavors to account for all the past phases in the history of the earth by the regular and ordinary action of the forces known to be at work in maintaining or modifying its present condition. The theory of meteoric agglomeration was first suggested by Julius Robert Mayer,* and was founded on the reflection that the great annual fall of meteoric masses upon the earth indicates the circulation or movement within our planetary space of a vast number of small bodies which must strike large bodies, like the sun, in numbers enormously exceeding those reaching the earth, the number being greater in proportion both to the masses and the surfaces of the larger bodies. These meteors, according to Mayer, are in a sense the fuel of the sun, and all bodies within the planetary system are subject to accretions, both of mass and temperature, in consequence of their collisions with them. Now, it is supposed that in astronomically primeval times the proportion of these meteoric masses to the masses of the large solar and planetary bodies may have been far greater than it is now—that, in fact, there may have been a time when the space now occupied by our planetary system presented the appearance of a swarm of such meteors of all sizes

* In his Beitraege zur Mechanik des Himmels (first published in 1848), Mechanik der Waerme, p. 157 *seq.*

and of all degrees and forms of consistency and aggregation, moving about at all rates of velocity, in all directions, and in orbits of every degree of eccentricity. These masses would be consolidated, and movements, both of rotation and revolution, would be generated in the bodies so formed by their collisions.

At this point the question obtrudes itself: how can a theory, which seeks to derive the orderly, symmetrical, and harmonious world as we know it from the wildest congestion of aboriginal differences and anomalies—from a spring-head of utter incongruity and confusion—be made to account for the regularities and coincidences whose simple and natural explanation was the conspicuous merit of the hypothesis of Laplace?

An answer to this question is sought, by the advocates of the new theory, in an appeal to a principle long since established by Laplace himself. This principle relates to the fact that, amid all the disturbances caused by the mutual attractions of the planetary bodies, there exists an invariable plane passing through the center of gravity of the whole system, about which these bodies perpetually oscillate with but slight deviations on either side. If on this invariable plane we project the areas described by the *radii vectores* of the several elements of mass in a given time, and multiply each mass into its respective area thus projected, the sum of the products is a maximum, and the rate of its increase is constant.* Such a plane exists, not only for the solar system, but for any system of bodies controlled solely by their mutual attractions. Now, it is evident that both the sum and the rate of its increase, of the products of the masses

* Cf. Laplace, Mécanique Céleste, 1ère partie, liv. ii, chap. vii. ("*Des inegalités séculaires des mouvemens célestes.*") The theory was first published in the Journal de l'Ecole Polytechnique, 1798.

into *the projections* of the areas described by their *radii vectores*, are always less than the sum and the rate of its increase of the products of the masses into the areas described by the *radii vectores* themselves, inasmuch as these *radii* (unless they are parallel to the plane) are shortened by their projection; and the difference between these two sums is in direct proportion to the deviations of the movements from the direction of the total increase, which direction, for purposes of reference, is taken as positive, the opposite direction being, of course, taken as negative. And whenever the several movements meet with resistance, some of the components of the velocities of the moving masses are necessarily destroyed, so that the difference in question is diminished and eventually annulled. When this has happened, the absolute values of the areas described by the *radii vectores* of the masses in a given time become equal to their maximum projections; in other words, their planes coincide with or become parallel to Laplace's invariable plane. From this follows the general principle that the movements of the bodies constituting any finite system, whatever be their original divergence of direction, tend (except in a very few special cases), by reason of any resistance to these movements, to become parallel to or coincident with an invariable plane.*

Before leaving this subject I may observe that the principle just stated, which admits of a further generalization, so as to assume this form—that all movements of the elements of a finite material system depending upon

* The possible exceptions to this law are, of course, those cases in which the components destroyed are exactly equal and opposite. The improbability of the occurrence of such cases is so great that Budde, who states the law substantially as I have stated it in the text (*l. c.*, p. 30), does not even allude to the possibility of an exception.

the mutual action of such elements tend, in consequence of any permanent interference with or determination of these movements from without, from irregularity and disorder to regularity and order—is, in my judgment, one of the most important in the whole range of theoretical physics. For the condition here assigned—that the internal movements of the system be subject to constant interference from without—is in fact inseparable from every material system, there being no such system which is at any time under the exclusive control of its own internal forces. There is, consequently, in every finite part of the world an ingenerate bias from irregularity to regularity, a natural bent from disorder to order, an inherent tendency from Chaos to Cosmos; and this tendency is the simple and direct consequence of the relativity of all material forms—of the fact that each finite whole is always a part of a still greater whole—in short, that the finite exists only on an ever-receding background of infinitude. It is possible even that this principle is more than coextensive with the sphere of physics, and that, to a certain extent, it may have its applications within the domains of those sciences which are ordinarily designated as historical. Although attempts at a transference of laws governing the interdependence of phenomena whose lines of connection are simple and easily traced (such as the movements of inorganic masses) to a class of phenomena whose relations are complicated and imperfectly understood (such as the phenomena of organic and vital action) are perilous in the extreme, and never to be made without a careful reference to the nature and ground of the analogies by which they are induced, it is nevertheless true that a great part of the progress which is now being made in the several departments of science is due to liberal exchanges, not only of results, but also of principles and methods.*

The theory of meteoric aggregation undertakes to grapple with still further elements of the general problem of explaining the actual features of our planetary system, as, for instance, the comparative inferiority of the sizes of the planets nearest the sun. The reasoning is something like this: Somewhere within the space comprising the various movements of the bodies, whose materials are in process of agglomeration, a mass will probably be formed which is preëminent above all the others. This—the nucleus of the future sun of the system—must gradually draw to its neighborhood the perihelia of all the moving meteoric masses or groups. In this region, therefore, the movements of all the bodies must have the greatest velocity; here the meteors must fly past each other with the greatest swiftness, and their approach and agglomeration must be most difficult—a circumstance which also prevents the rapid growth of bodies in this region after their inchoate formation. Near the confines of the system, on the contrary, where the movements of the meteors are sluggish, the conditions for the congestion of large masses are comparatively favorable. Similarly, a rough account is given of the fact that the densities of the planets are

* Instances of the application of dynamical and, generally, of physical laws, not only to vital, but also to psychological, action are afforded by the recent discussion, by Avenarius, of the evolution of thought in conformity to the principle of least action (Die Philosophie als Denken der Welt gemaess dem Princip des Kleinsten Kraftmaasses, Leipzig, 1876), and the previous discussion, by Schleicher, of the evolution of language in the light of the doctrine of natural selection—which, it may be said parenthetically, is not without analogy to the principle discussed in the text—(Die Darwin'sche Theorie und die Sprachwissenschaft, Weimar, 1863).

generally in the inverse ratio to their sizes. A larger body attracts a meteor with greater intensity than a smaller one; its growth is, therefore, marked by more violent collisions productive of a higher temperature and a corresponding expansion.

It is not my purpose to discuss the merits of this theory in detail, or to express an opinion as to its soundness and sufficiency; but it is proper to say that it appears to me to stand in favorable contrast to the nebular hypothesis precisely by reason of the absence of some of the characteristics to which the general plausibility of this latter hypothesis is due. The nebular hypothesis found ready and almost enthusiastic acceptance, not so much on physical as on metaphysical grounds. The proneness to derive the Multiple from the absolutely Simple, the Various from the absolutely Uniform, has its root in the second of the great structural fallacies which I have discussed in the ninth chapter—in the assumption that the abstract result of a generalization, i. e., a general concept, may be made available as a starting-point for the evolution of the particular things subsumed under it. The enthusiasm for the nebular hypothesis was, in this respect, an ontological survival. And in another respect it was even more than that—it was a recrement of ancient traditions about the origin of the universe from Nothing. The original mist of the nebular hypothesis is assumed to be of extreme tenuity—of a density less than the one hundred thousandth part of hydrogen, the lightest gaseous body known to the chemist. By reason of this æthereal subtilty it was readily substituted, in the conceptions of the popular mind, for the old void from which the world was said to have emerged, and, in the imaginations of those who look upon matter as a sort of inspissation of Mind for the universal antemundane impersonal Spirit. It thus conformed to the assumption that, on any hypothesis respecting the mode of the world's formation, it must "in the beginning" have been "without form and void," and at the same time satisfied the mystic yearnings after the Ethereal and "Spiritualistic," which is the special distinction of that large class of philosophers whose philosophy begins where clear thinking ends.

CHAPTER XVI.

CONCLUSION.

The considerations presented in the preceding pages lead to the conclusion that the atomo-mechanical theory is not, and can not be, the true basis of modern physics. On proper examination, this theory appears to be not only, as is generally conceded, incompetent to account for the phenomena of organic life, but it proves to be equally incompetent to serve as an explanation of the most ordinary cases of inorganic physical action. And the claim that, in contradistinction to metaphysical theories, it resorts to no assumptions, and operates with no elements save the data of sensible experience, is found to be wholly inadmissible. In announcing this conclusion it is necessary, however, to guard against two fundamental misconceptions. In the first place, the denial of the theory of the atomic constitution of matter, as it is generally held by physicists and chemists, involves no assertion respecting the real constitution of bodies—of chemical elements or compounds—and certainly does not imply the metaphysical thesis of the absolute continuity of matter. What is the actual constitution of particular bodies is a question to be determined in each case by experiment and observation. There is, no doubt, a large class of bodies whose constitution is molecular; but from this it does not follow that the molecules composing them are primordial, unchangeable units, existing independently and in advance of all physical action, and therefore absolutely exempt from change. On empirical grounds the inference, from the molecular structure of a body, of the permanent existence of absolutely immutable and indestructible atoms or molecules is as irrational as would be the assertion that primordially, and in advance of the formation of organic bodies, there existed an indefinite number of elementary cells, because all organic bodies are of cellular structure.

In the second place, dissent from the proposition that all physical action is mechanical in the sense of being a transference of motion between distinct masses by collision or impact is not to be construed as a doubt respecting the constancy of physical laws or the universality of their application. What is denied is, not the general dominance of the law of physical causation, but the doctrine that the only form of such causation is the transference of motion by the impact of masses which, in themselves, are absolutely inert. If physical action in conformity with constant and uniform law is designated as mechanical, then all physical action is undoubtedly mechanical.

It may be said that physical action is utterly indeterminable except on the supposition of the atomic or molecular constitution of matter. This is true only in the sense that we are unable to deal with forms of physical action otherwise than by considering them as modes of interaction between distinct physical terms. Physical action can not be subjected to quantitative determination without a logical insulation of the conceptual elements of matter, and without ultimate reference to conceptual constants of mass and energy. All discursive reasoning depends upon the formation of concepts, upon the intellectual segregation and grouping of attributes—in other words, upon the consideration of phenomena under particular aspects. In this sense the steps to scientific as well as other knowledge consist in a series of logical fictions which are as legitimate as they are indispensable in the operations of thought, but whose relations to the phenomena whereof they are the partial and not unfrequently merely symbolical representations must never be lost sight of. When the old Greek sought to determine the properties of the circle, he began by constructing a polygon whose sides he subdivided until they were supposed to become infinitely small; and in his view every line of definite extent and form—i. e., every line which could become the subject of mathematical investigation—was composed of an infinite number of infinitely small straight lines. But, he speedily found that, while this fiction enabled him to deduce a rule for calculating the area of the circle and otherwise to determine a number of its properties, nevertheless the circle and its rectilinear diameter were fundamentally incommensurable, and the quadrature of the circle was impossible. The modern analyst similarly determines the locus of a curve by the relation of small increments of coördinates arbitrarily established; but he is well aware that the curve itself has nothing to do with this arbitrary representation, and he very emphatically asserts the continuity of the curve by differentiating, or passing to the limit of, his increments—at the same time transforming his coördinates by changing their origin or their inclination, or even their system, from bilinears to polars, whenever he finds it convenient, without dreaming that thereby he is in the least affecting the nature of the curve whose properties are under discussion. The astronomer, in calculating the attraction of a homogeneous sphere upon a material point, begins by assuming the atomic or molecular constitution of the attracting sphere, establishing a series of finite differences as one of the terms of his equation; but thereupon he takes the series to be infinite and the differences to be infinitely small, and very effectually dismantles the molecular scaffolding by integrating instead of effecting a summation of a series of finite differences. Observe: the astronomer begins with two fictions—the fiction of a "material point" (which is, in truth, a contradiction in terms), so as to insulate the attractive force and treat it as proceeding from the sphere alone, and the fiction of the finite differences representing the molecular constitution of the sphere; but the validity of his result depends upon the eventual rescission of these fictions and the rehabilitation of the fact. In like manner the chemist represents the proportions of weight, in which substances combine, as atoms of definite weight, and the resulting compounds as definite groups of such atoms; and this mythical coinage has been serviceable in many ways. But, apart from the circumstance that the symbols have become wholly inadequate to the proper representation of the facts, it is important to bear in mind always that the symbol is not the fact. Newton derived many of the leading optical laws from his corpuscular theory of light and from the hypothesis of "fits of easy transmission and reflection." His theory for a time served a good purpose; but it proved, after all, to be but a convenient mode of symbolizing the phenomena with which he was familiar, and had to be discarded when the phenomenon of interference was observed. In 1824 Sady Carnot deduced the law of thermic action which still bears his name from an hypothesis respecting the nature of heat (supposed by him, as by nearly all the physicists of his time, to be imponderable matter), which is now known, or universally believed, to be erroneous. For certain purposes, such as the mathematical determination of gaseous pressure and expansion, thermic phenomena find a convenient representation in the hypothesis that a gaseous body is a group of atoms or molecules in a state of incessant motion. Some of the properties of gases have been successfully deduced, by Clausius and others, from formulæ founded upon this hypothesis, and Maxwell has even succeeded in predicting the phenomenon of the gradual cessation of the oscillatory movement of a disk, suspended between two other disks, in consequence of the friction of a gaseous medium, whatever be the degree of its tenuity, and this prediction has since been experimentally verified; but neither Clausius's formulæ nor Maxwell's experiments are conclusive as to the real nature of a gas. That no valid inference respecting the real constitution of bodies and the true nature of physical action can be drawn from the forms in which it is found necessary or convenient to represent or to conceive them, is illustrated by the fact that we habitually resort, not only in ordinary thought and speech, but also for purposes of scientific discussion, to modes of representing natural phenomena which are founded upon views and hypotheses long since discarded as untenable. Just as we think and speak familiarly of the motions of the sun and stars in terms of the old geocentric doctrine, although no one in our day doubts the truth of the heliocentric theory, so also the modern astronomer would find it difficult to dispense with geocentric fictions in subjecting these motions to mathematical computation.

Even the old epicycles survive in some of the analytical formulæ, by means of which that computation is effected.

The progress of modern theoretical physics consists in the gradual reduction of the various forms of physical action to the principle of the conservation of energy. For purposes of didactic exposition of this principle we resort to the fiction of systems of molecules or particles, whose motions are simple functions of the distances between them. But, as we have seen, the conflict of this fiction with the facts of experience emerges at once, when we undertake to establish an absolute disjunction between the molecules and their motions. The conservation of energy would be impossible, if the ultimate constituents of a material system were in themselves absolutely inert. And the same thing is strikingly exhibited in the recent attempts to extend the principle of the conservation of energy to the phenomena of chemical action. These attempts have been prompted by the observation that all chemical action depends upon, or, at least, is attended with, the absorption or liberation of heat, and that the amount of heat absorbed or liberated is the measure of such action. The determination of chemical phenomena by means of their thermic incidents, which has until recently been known as thermochemistry, and has been treated as a comparatively insignificant part of chemical science, is now coming to be regarded as the true basis of theoretical chemistry. The principles of this new science have already been systematized, to some extent, in several distinct treatises, among which may be mentioned Mohr's "Mechanical Theory of Chemical Affinity," * Naumann's

* Friedrich Mohr, Mechanische Theorie der chemischen Affinitaet, Braunschweig, 1868.

"Thermo-Chemistry,"* and Berthelot's "Chemical Mechanics founded on Thermo-Chemistry."†

The importance of the part which heat performs in chemical transformations was first distinctly realized upon the announcement, by Dulong and Petit in 1819, of the empirical law, that the specific heats of the elements are inversely proportional to their atomic weights, or, as it is commonly expressed in the language of the atomic theory, that the atoms of all elementary bodies have the same specific heat. Although there are apparent exceptions to this law (as in the cases of carbon, boron, and silicon), it holds good in so many cases that there is hope of an explanation of these exceptions on grounds on which they will utimately prove to be confirmations of the law; indeed, some progress in this direction has already been made. And Neumann, Regnault, and Kopp have shown that the law applies not only to elements, but also to compounds, it appearing that the specific heat of a compound is the sum of the specific heats of its component elements.

Dulong and Petit's law, if it were universally valid, would lead to a remarkable law of chemical combination. For, it is obviously identical with the proposition, that chemical elements combine only in so far as they experience the same elevation of temperature in the act of combination. It is not improbable that, if the true relation of the temperature of a body to its total physical and chemical energy were thoroughly understood, this law would become one of the cardinal principles of theoretical chemistry.

* Dr. Alexander Naumann, Grundriss der Thermochemie, Braunschweig, 1869.

† M. Berthelot, Essai de Mécanique Chimique fondée sur la Thermochimie, Paris, 1879.

The next noteworthy result of thermo-chemical research was the discovery that the nature of the chemical reactions between different substances depends upon the relations between the specific energies of the reagents as determined by the quantities of heat evolved or involved in the progress of these reactions. It was found that there are certain elements—oxygen and hydrogen, for example—which combine readily, and, under proper conditions, spontaneously, the combination (as Berthelot expresses it) taking place directly, without the aid of extrinsic energy, and being attended with the evolution of light, or heat, or both. Such combinations are termed by M. Berthelot, *exothermic.* They result in the formation of compounds, which can not be resolved again into their original elements without a restoration of the amount of energy lost in the combination. On the other hand, there are cases of *endothermic* combination in which conversely the composition of the elements is attended with an absorption, and the decomposition of the resulting compound with a liberation, of heat. The combination of carbon and sulphur, for instance, is endothermic. Carbonic disulphide is formed by passing vaporous sulphur over red-hot charcoal; the union of carbon and sulphur is possible only on condition of the continuous supply, during the progress of the union, of heat, which is given out again when the disulphide is resolved into its elements. The facts here referred to are explained, by the modern chemist, on the theory that chemical affinity is transformed heat, both heat and affinity being forms of energy; that in the cases of exothermic combination the sum of the specific energies of the component elements exceeds the specific energy of the compound formed, while in endothermic combinations the specific energy of the compound is greater than the aggregated specific energies of the components. And it has been shown that, whenever we trace a number of elements or compounds through a series of chemical reactions, the total amount of energy (appearing, before absorption or after liberation, in the form of heat), which is liberated or absorbed, is exactly equal to the difference between the specific energies of the initial and those of the terminal compounds or elements. It is to be observed that this rule applies, not only to cases of composition and decomposition, so called, but likewise to cases of allotropy and polymerism, inasmuch as allotropic forms of elements and isomeric forms of compounds are found to be convertible into each other by the addition or withdrawal of definite amounts of heat.

A third result of the study of the thermic condition of elements and compounds is the establishment of the remarkable principle that the passage of any body or system of bodies from a condition of a lesser to one of greater stability is always attended with evolution of heat, "whether" (in the language of Odling) "such change be what is commonly called combination, or what is called decomposition"; and that all chemical action which takes place without the intervention of extrinsic energy tends to the production of a body or bodies whose formation liberates the largest amount of heat.*

* A sort of anticipation of this principle is found in one of the well-known laws announced in the early part of this century by M. Berthollet, in his "Statique Chimique"—in the law that, whenever two soluble salts are mixed in solution, they decompose each other, if the resulting compound, or mixture of compounds, is insoluble or less soluble than the salts mixed. The bearing of this law upon the principle, stated in the text, of the maximum evolution of heat, will be understood upon reference to the fact that, generally speaking, the solubility of substances is increased by

This brief outline sufficiently indicates the facts and generalizations upon which it is proposed to found the new theory of "Chemical Mechanics." Little use has, thus far, been made of the law of Dulong and Petit; but the other results of experimental induction in the field of thermo-chemistry are summarized by M. Berthelot in the introduction to his work,* as follows:

"1. *Principle of Molecular Work.*—The quantity of heat disengaged in any reaction whatever is a measure of the amount of chemical and physical work performed in such reaction.

"2. *Principle of the Calorific Equivalence of Chemical Transformations.*—If a system of simple or compound bodies, taken under determinate conditions, undergoes physical or chemical changes capable of bringing it to a new state, without producing any mechanical effect outside of the system, the amount of heat liberated or absorbed by the effect of these changes depends solely on the initial and final states of the system; it is the same, whatever be the nature and sequence of the intermediate states.

"3. *Principle of Maximum Work.*—All chemical change effected without the intervention of extraneous energy tends to the production of that body or system of bodies which liberates the largest amount of heat."

This third principle, as Berthelot observes, may also be stated in the form that "all chemical reaction susceptible of being effected without the concurrence of preliminary work and without the intervention of ex-

the application of heat. Berthollet's law, however, is subject to exceptions; there are cases in which soluble bases are replaced by insoluble bases, the result, nevertheless, being the formation of soluble salts.

* "Mécanique Chimique," pp. xxviii, xxix.

trinsic energy will necessarily take place whenever it leads to the evolution of heat."

The relation of these propositions to the doctrine of the conservation of energy is apparent. They are obviously applications, to the phenomena of chemical transformation, of the two leading principles which that doctrine embraces, the first and second propositions of Berthelot representing the principle of the correlation, equivalence, and mutual convertibility of the several forms of energy, and the third that of the tendency of all energy to dissipation.

The study of chemical changes in the light of the doctrine of the conservation of energy exhibits these changes under an entirely new aspect. It shows that the question as to the possibility of a chemical "composition," or "decomposition," is as much a question of the definite proportionality of energies as of the definite proportionality of masses; that each element as well as each compound embodies a distinct and invariable amount of energy as well as a distinct and invariable quantity of "matter" (i. e., mass), and that this energy is as constitutive, and as essential a part, of the existence of such element or compound as its weight.

And here the question arises: How is all this to be interpreted, by the aid of the ordinary laws of motion and of mechanical principles generally, in conformity with the assumption that all the phenomena of chemical transformation are reducible to motions of absolutely inert atoms or elements of mass? For that is the assumption which lies at the base of the new theory of chemical mechanics. Naumann declares in express terms, both in one of the first and in the very last of the sentences of his book that "chemistry in its ultimate form must be atomic mechanics." * And Berthelot, though he avoids the use of the word *atoms*, no less explicitly asserts that two data suffice to explain the multiformity of chemical substances: the masses of the elementary particles and the nature of their motion.†

The explanation of chemical phenomena by the theory of chemical mechanics is to be effected, then, by reducing them to terms of mass and motion. On what mechanical principles is this reduction possible? The fundamental fact to be accounted for is the conversion of heat into chemical energy. But this conversion implies, not only a change of one kind of motion into another, but also a confinement of a definite amount of this motion to or within a definite mass. According to the mechanical theory, heat, in the form at least in which it is generally supplied to gaseous bodies in process of chemical transformation, consists in rectilinear atomic or molecular motions of all conceivable velocities and directions. The extent of these motions is limited solely by the encounters of the moving masses. By these encounters the range, the velocity, and the direction of the excursion of every atom or molecule are incessantly changed. And, whatever may be the nature of that form of motion which we call chemical energy, we know at least that a definite and invariable amount of it belongs to a definite mass or

* "*Die Chemie in der fuer sie zu erstrebenden Gestaltung muss sein eine Mechanik der Atome.*" Thermochemie, p. 150.

† "*La matière multiforme dont la chimie étudie la diversité obéit aux lois d'une mécanique commune. . . . Au point de vue mécanique, deux données fondamentales caractérisent cette diversité en apparence indéfinie des substances chimiques, savoir : la masse des particules élémentaires, c'est-à-dire leur équivalent, et la nature de leurs mouvements. La connaissance de ces deux données doit suffir pour tout expliquer.*" Mécanique Chimique, tome ii, p. 757.

number of atoms of any given substance. Whenever heat, therefore, is converted into chemical energy, the motion above-described must, of necessity, be so modified that a definite amount of it is brought into some sort of synthesis or union with a definite number of particles. But that is certainly impossible if the particles are mere inert masses, whose motions are determined solely by the impact of other masses, as the mechanical theory assumes. The specialization or individualization of motion, which is required, can be accounted for in no other way than by attributing to the masses themselves some inherent coercive power. Even if an individualization of heat-movements could result mechanically from the collision of inert particles —by the conversion of rectilinear into rotatory motion, for instance, as a consequence of oblique impacts—there would still remain the impossibility of accounting for the fact that such conversion invariably ceased at the precise moment when each atom or molecule had been supplied with its due amount of energy.

In view of all this it is strange to read in the writings of distinguished physicists sentences like these: "The only real things in the physical universe are matter and energy, and of these matter is simply passive," * and, "We see that, whereas (to our present knowledge at least) matter is always the same, though it may be masked in various combinations, energy is constantly changing the form in which it presents itself. The one is like the eternal, unchangeable Fate or *Necessitas* of the ancients; the other is Proteus himself in the variety and rapidity of its transformations." †

There is little doubt that the principle of the con-

* The Unseen Universe, § 104. † *Ib.*, § 103.

servation of energy will prove to be the great theoretical solvent of chemical as well as of physical phenomena; but thus far, at least, the endeavor to express the laws of chemical action in terms of mass and motion or kinetic energy has been as abortive in chemistry as in physics. To what extent it may be possible, hereafter, to bring the phenomena of chemical action within the dominion of the mechanical laws controlling the interaction of solids, it is difficult to determine. There are, however, several well-known facts which appear to indicate that, whatever be the nature of chemical energy, it can hardly result from the impact of solid particles. The chemical energies of the elements are proportional neither to their masses as measured by their weights, nor to their volumes; and their mechanical equivalents are so enormous as to seem out of all analogy to ordinary mechanical action. In 1856 W. Weber and R. Kohlrausch published the results of a series of investigations by which they had sought to arrive at a mechan ical measure for the intensity of a galvanic current. They applied these results to the electrolytic decomposition of water, so as to determine the energy represented in the chemical union of hydrogen and oxygen. And they announced their conclusion in the following words*: "If all the particles of hydrogen in one milligramme of water contained in a column of the length of one millimetre were attached to a string, the particles of oxygen being attached to another string, each string would have to be under a tension, in a direction opposite to that of the other, of 2,956 cwt. (147,830 kilogrammes), in order to effect a decomposition of the water with a velocity of one milligramme per second" And, looking to the equivalents of chemical energy in

* Pogg. Ann., vol. xcix, p. 24.

terms of units of heat, it has been found that the combination of a gramme of hydrogen with 35·5 grammes of chlorine, so as to form 36·5 grammes of hydrogen chloride, is attended with the liberation of an amount of heat by which the temperature of 24 kilogrammes of water would be raised one degree; inasmuch, therefore, as the heat required to raise one kilogramme of water one degree is mechanically equivalent to 425 kilogram-metres, the formation of 36·5 grammes of hydrogen chloride gives rise to a power by which a weight of 10,000 kilogrammes can be raised to the height of one metre in a second.

INDEX.

16

PRINTED BY
SPOTTISWOODE AND CO., NEW-STREET SQUARE
LONDON

Made in the USA
Monee, IL
04 November 2026

67391589R00203